Polynomials of One Variable

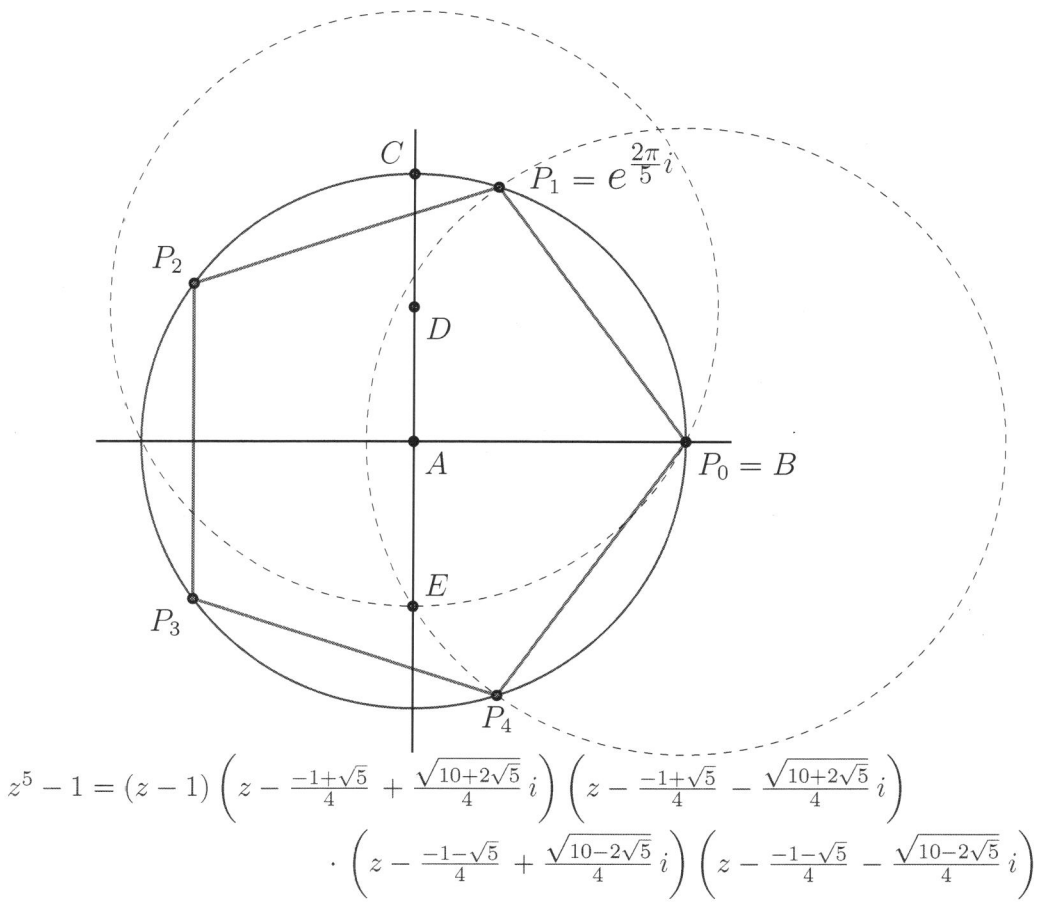

$$z^5 - 1 = (z-1)\left(z - \tfrac{-1+\sqrt{5}}{4} + \tfrac{\sqrt{10+2\sqrt{5}}}{4}i\right)\left(z - \tfrac{-1+\sqrt{5}}{4} - \tfrac{\sqrt{10+2\sqrt{5}}}{4}i\right)$$
$$\cdot \left(z - \tfrac{-1-\sqrt{5}}{4} + \tfrac{\sqrt{10-2\sqrt{5}}}{4}i\right)\left(z - \tfrac{-1-\sqrt{5}}{4} - \tfrac{\sqrt{10-2\sqrt{5}}}{4}i\right)$$

Polynomials of One Variable:
The Classical Theory of Equations

Chris K. Caldwell
University of Tennessee at Martin
Martin, TN 38238
caldwell@utm.edu

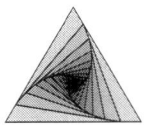 Ossuary Press (March 10, 2016)

© 2015 Chris K. Caldwell

All rights reserved.

No part of this publication may be reproduced or transmitted in any form or by any means, electronic or mechanical, including photocopy, recording, or any information and retrieval system without permission in writing from the authors.

ISBN-13: 978-1514800706

Preface

THIS BOOK GREW OUT OF A NEED. Our university occasionally teaches cohorts of secondary school teachers—hoping to enhance their mathematical skills and facilitate the sharing of teaching techniques. We sought a course that would build on the algebra that they regularly teach, and from that basis refresh and expand their knowledge of algebra, complex numbers and geometry. We also hoped to deepen their appreciation of the interplay between these subjects. It is in this interaction between areas that mathematics truly comes alive!

The minimal knowledge prerequisite is college algebra, although in a few sections we assume a minimal knowledge of calculus. Also prerequisite is the mathematical sophistication hopefully acquired by students before taking upper division courses.

The numerous exercises range from simple computations and practice to theorem generation, generalization and proof. This variety provides great flexibility, allowing this text to be used in courses taught at many levels of sophistication. Exercises are included which explore the overlap of the theory of equations with computer programming, geometry programs, ring theory (abstract algebra) and number theory. These are marked with ▨ , ▷ , (A) and (N) respectively. (We have used the more difficult problems and those which require information from other courses to teach this class at a beginning graduate level.)

Chapter two introduces the complex numbers and their usual representations. This is done quickly in the hope that it is a review. Sections are devoted to both Euler's Formula and DeMoivre's Theorem. This chapter concludes with a discussion on the nth roots of complex numbers.

Chapter three introduces polynomials, polynomial division and factorization. We discuss the Remainder and Factor Theorems along with their usual application using synthetic division. (We leave long division with detached coefficients to Appendix B.2). Chapter three concludes with a discussion of greatest common divisors and the Euclidean algorithm for polynomials.

In chapter four we start our study with the zeros of polynomials, beginning with the Fundamental Theorem of Algebra which shows that a polynomial of degree n has n zeros. (A proof of this theorem is sketched in the Appendix A.1.) We glance at the complex zeros of real polynomials, and demonstrate how to find multiple zeros using the derivative. As a partial inverse to the problem of finding the zeros of a given polynomial, we conclude chapter four with Lagrange's Interpolation Formula—a polynomial that takes specified values at given points.

In chapter five we introduce the Rational Zero Theorem as a method to find rational and integral zeros of polynomials. We then give ways to improve on this theorem and present Eisenstein's Irreducibility Criterion. We next discuss a relatively easy to solve class of polynomials: the reciprocal polynomials. Chapter five concludes with a discussion of Lill's amazing geometric solution of polynomial equations—not the most practical of approached, but fun fusion of geometry and algebra.

Chapter six studies the relationships between the zeros of polynomials and their coefficients. We use these relationships to transform and solve polynomial equations. Chapter six concludes with the Fundamental Theorem on Symmetric Functions, giving Newton's Identities as one of the examples.

Chapter seven presents the algebraic solutions to polynomials of degrees one through four. We will also discuss why there are no general algebraic solutions for polynomials with higher degrees.

In chapter eight, we use what we have learned to solve the classical ruler and compass construction problems: proving that we can not trisect the general angle, duplicate the cube, or construct most regular polygons. We begin by carefully defining the geometric concept of constructability. We then use the relationship between the cartesian plane and the complex numbers to translate the question of the constructability of a point, into the question of whether the corresponding complex number belongs to a particular set (the set of constructible numbers). We finally reduce this membership question to one of knowing the degree of a certain polynomial. After this, we can answer the classic questions by simply viewing the appropriate polynomials. All mathematics students could benefit by seeing this triumph of modern mathematics over classical geometry problems. Chapter eight ends by discussing how we might bend the rules to solve these classically unsolvable problems. For example we show how origami can be used to address these same problems—and to solve many that we not solvable by ruler and compass alone.

Chapters nine and ten return us to zero finding, but in a more general context. In chapter nine we present methods to count, separate and bound the real zeros of polynomials. These included Descartes' Rule and Sturm's Theorem. In chapter ten we present numerical methods to approximate zeros beginning with bisection and Newton's method.

Finally it seems natural to study that set of numbers which are the zero of some polynomials: the algebraic numbers. We do this in chapter eleven. In this chapter we also study the rational, irrational and transcendental numbers.

This text also has hundreds of exercises ranging from the most basic applications to those that will require exploration and proof. One reason for this wide variety of exercise levels (and the increased difficulty of chapter eleven) is so that we can teach this course at a variety of different levels. Appendix C contains answers to selected problems and the main text provides over 100 completely worked out examples.

More could have been done with computers in this text. However our main purpose is to build the student's skill sets and confidence. One of the best ways to do this is to work things out for yourself, by hand and by mind. It is easy to delegate too much of the work to machine and calculators, leaving the user dazzled but uninformed. We must learn to master, not be mastered by, our technology.

After studying this material, we hope that the reader is encouraged to question and explore further on their own. New discoveries in mathematics are free for the finding, and we all have the necessary equipment to begin the search: a pencil, some paper and the minds we were each born with.

I would like to thank Ann Gardner and Thomas Eskew for their help while producing this text. Dana Stanfill kindly wrote the solutions to the exercises found in Appendix C. Stephanie Kolitsch especially has helped make this a far better book—she is a true hero of mine! Any errors or omissions that remain are the fault of the author alone.

A comment on the exercises

You cannot learn by osmosis—you should work as many of the exercises as possible. Often additional definitions and results are presented in the exercises, so any problems that are not attempted should at least be read.

The Theory of Equations overlaps many areas of mathematics. Some of this overlap is developed in the problems and these problems may involve extra knowledge. These problems may be indicated with a combination of the following symbols.

(*)	difficult	May require extra time, thought or effort.
(**)	more difficult	Requires extra time, thought and effort.
(***)	very difficult	A good independent study problem.
(A)	ring theory	Requires some group, ring or field theory.
(N)	number theory	Easier with some knowledge of number theory.
	GeoGebra	Easier with a system such as GeoGebra.
	computer	Requires knowledge of a programming language.

Selected answers may be found in Appendix C. Many of the problems also include hints. Remember there is almost always more than one way to solve a problem. Whenever possible, give a better solution than the solution suggested by the hint (or found in the back of the text).

Contents

Preface		v
1	**Introduction**	**3**
2	**Complex Numbers**	**7**
	2.1 Definition of the Complex Numbers	7
	2.2 The Complex Plane	12
	2.3 The Polar Form of Complex Numbers	16
	2.4 Euler's Formula	21
	2.5 De Moivre's Theorem	24
	2.6 Roots of Complex Numbers	26
3	**Polynomials**	**33**
	3.1 Basic Definitions	33
	3.2 The Remainder and Factor Theorem	35
	3.3 Synthetic Division	38
	3.4 The Greatest Common Divisor	42
4	**Zeros of Polynomials**	**51**
	4.1 Polynomial Equations	51
	4.2 The Fundamental Theorem of Algebra	54
	4.3 Imaginary Zeros	57
	4.4 Finding Multiple Zeros	60
	4.5 Lagrange's Interpolation Formula	62
5	**Elementary Techniques for Finding Zeros**	**67**
	5.1 Rational Zeros	67
	5.2 Integral Zeros of Polynomials	71
	5.3 Eisenstein's Criterion	75
	5.4 Reciprocal Polynomials	80
	5.5 Lill's Method for Solving Polynomials	85
6	**Relationships Between the Zeros and the Coefficients**	**95**
	6.1 The Coefficients	95

	6.2	Transforming the Zeros of Polynomials I	100
	6.3	Transforming the Zeros of Polynomials II	105
	6.4	Symmetric Functions	108
	6.5	Newton's Identities	113

7 Algebraic Solutions of Polynomials — 119
7.1 General Polynomial Equations of Degree n — 119
7.2 Solution of the General Cubic — 122
7.3 Solution of the Cubic: Irreducible Case — 127
7.4 Solution of the General Quartic — 130

8 Geometric Constructions — 137
8.1 Squaring the Circle — 137
8.2 The Geometric Concept — 140
8.3 The Analytic Concept — 146
8.4 The Algebraic Concept — 150
8.5 Regular Polygons — 154
8.6 Closing Comments — 163

9 Separating the Real Zeros of Polynomials — 175
9.1 Separating the Zeros — 175
9.2 Descartes' Rule of Signs — 179
9.3 Bounds for the Zeros of Polynomials — 183
9.4 Bounds for Real Zeros — 186
9.5 Sturm's Theorem — 193

10 Approximating Real Zeros: Numerical Methods — 201
10.1 Sure and Simple: Bisection — 201
10.2 Faster But Not Sure: Newton-Raphson — 207
10.3 A Gamble: Simple Iteration — 214

11 Numbers: Rational, Irrational, ... — 221
11.1 Rational or Irrational? — 221
11.2 Algebraic or Transcendental? — 231
11.3 Measuring Irrationality — 235

Appendices — 241

A Delayed Proofs — 243
A.1 The Fundamental Theorem of Algebra — 243
A.2 Proof of Sturm's Theorem — 246
A.3 Bundan-Fourier and Descartes' Rule — 248

B Useful Results   **253**
    B.1 The Resultant of Two Polynomials 253
    B.2 Division with Detached Coefficients 256

C Selected Answers   **259**

Bibliography   **265**

Index   **269**

Introduction

A mathematician, like a painter or a poet, is a maker of patterns. ... The mathematician's patterns ... must be beautiful; the ideas must fit together in a harmonious way. Beauty is the first test: There is no permanent place in the world for ugly mathematics.

G. H. Hardy

I had a feeling once about Mathematics—that I saw it all. Depth beyond depth was revealed to me—the Byss and the Abyss. I saw—as one might see the transit of Venus or even the Lord Mayor's show—a quantity passing through infinity and changing its sign from plus to minus. I saw exactly why it happened and why the tergiversation was inevitable—but it was after dinner and I let it go.

Winston Churchill

Chapter 1

Introduction

IN THE SIXTEENTH AND SEVENTEENTH CENTURIES mathematics came into full bloom and many expected the new arts of algebra and physics to answer all questions. Mathematicians and amateurs alike participated in public problem solving contests—displaying their prowess with polynomials and mental computation. In fact, the correctness of solutions and the authorship of ideas was often decided by public debate—without modern mathematics' (usual) calm objectivity, and often without rigorous logic. This is where the theory of equations was born.

The theory of equations is the study of polynomial equations. The most studied equations were polynomial equations two hundred years ago. In this book we get to rediscover the relationships between the coefficients and the zeros of polynomials, between multiple zeros and the derivative, between the powers of the number e and the trigonometric functions. We read here, as simple formulas, the solutions over which careers began and lives were shattered many years ago.

One quickly discovers that when using only the real numbers (those represented on a number line) we cannot solve most polynomial equations, even equations as innocuous looking as $x^2 + 1 = 0$. So the study of polynomials gave birth to the complex numbers—the number system based on the square root of negative one. These numbers have marvelous properties which we will study and enjoy in the next chapter, developing tools to be applied throughout the text. In the theory of equations we find the roots of the modern field of complex variables, now invaluable to much of physics, electronics and engineering.

We are all familiar with the quadratic formula for solving quadratic polynomials, but do you know that there are also formulae for solving cubic and quartic polynomials (polynomials of degree three and four)? Do you know that there are no (and can be no) such formulae for general polynomials of any higher degree? The study of such algebraic solutions is an important part of the theory of equations, the part that gave birth to the modern field called abstract algebra.

When you cannot solve polynomial equations of higher degrees exactly, or when

1. Introduction

these solutions are impractically difficult, then you look for ways to approximate the solutions. So it is only natural that the theory of equations is full of numerical techniques for manipulating, transforming and solving equations. Here again the theory of equations gave birth to another child, the modern field of numerical analysis.

Certainly the theory of equations cannot claim to have birthed geometry, but it has been applied to geometry with great success. Several hundred years before Christ the Greek geometers asked if "given a circle, can we construct a square with the same area as the circle using only a ruler and compass?" That is, they asked, can you "square the circle?" They also wondered if "given an angle, can we divide it into three equal parts" (can you "trisect any angle")? The theory of equations was instrumental in showing that the answer is no to both of these questions. Later in this text we prove this, and prove many related results. This proof provides one of the greatest examples in undergraduate mathematics of how mathematics works: using careful definition, abstraction, generalization and simple logic. It is a tribute to the beauty and appeal of classical geometry that even though these geometrical constructions were proved impossible over two centuries ago, amateur mathematicians are still trying to do them—and still claiming (false) success!

However, the theory of equations now lies dying, dissected by its own success. Its main parts remain in the fields mentioned above, but it is sadly neglected as a whole—especially as a course in today's universities. It would be incorrect to blame the death totally on success. Computers have rearranged the undergraduate mathematics curriculum, and the theory of equations is but one of the casualties. After the *relatively* recent growth of abstract algebra as a central undergraduate course, the theory of equations' last stronghold was in its numerical methods. But calculators have replaced the need to take cube roots by hand. When did you last approximate a square root by hand? Who needs specialized algorithms that give good approximations in just twenty steps by hand, when generalized (computer) algorithms work in twenty thousand steps—and these twenty thousand steps take less than a fraction of a second on your handheld device? We no longer need to understand polynomials when they can be solved by brute force. Thus computers removed the last stronghold of the theory of equations, leaving numerical methods as a computer science course, often ignorant of its own birthright.

Despite this, and perhaps even because of it, the theory of equations is an invaluable "survey" course for the secondary mathematics teacher and an excellent refresher for those that have been in the K12 classroom a decade or two. The solution of the classical construction problems are an unparalleled example of mathematics from which all students can benefit. A deeper understanding of polynomials, which are so central to high school algebra, can inform our teaching and expand our understanding of even the most basic algorithms. We have successfully taught this course several times as a training and retraining course for many of the best high school teachers in our area.

The Complex Numbers

The Divine Spirit found a sublime outlet in that wonder of analysis, the potent of the ideal, that mean between being and not-being, which we call the imaginary [square] root of negative unity.

Liebniz

It would be naturally expected that the discovery of imaginaries, which seem nearer to madness than to logic and which, in fact, has illuminated the whole of mathematical science, would come from such a man whose adventurous lie was not always commendable from the moral point of view, and who from childhood suffered from fantastic hallucinations...

Jacques Hadamard speaking about Cardan

Chapter 2

Complex Numbers

Our goals in this text include understanding both polynomials and key parts of geometry. Both can be best understood with the help of complex numbers. For example, in a later chapter the vertices of regular polygons will be viewed as the complex zeros of simple polynomials, and from these polynomial we will be able to determine which regular polygons can be geometrically constructed with a ruler and compass. But this is getting far ahead of ourselves!

In this chapter we will start with what you should recall from "college algebra" (or perhaps high-school algebra). We will take this a little further by presenting the polar form of complex numbers. The polar form simplifies the multiplication and division of complex numbers and connects algebra to trigonometry. This is exemplified by Euler's Formula and De Moivre's Theorem. The latter makes it easy to derive certain trigonometric identities (and is the way I recalled double and triple angle formulas while in high-school).

2.1 Definition of the Complex Numbers

THE NUMBERS WE USE in elementary algebra and calculus are called the **real numbers** (denoted by $\mathbb{R}$). They are often represented by the points on a line called the "real number line." Since the square of a real number is never negative,

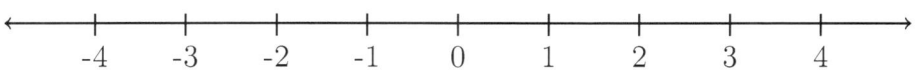

Figure 2.1: The Real Number Line $\mathbb{R}$

there is no real number x for which $x^2 = -1$. For this reason we need to expand the set of real numbers by adding more elements. In particular we add a number i

2. COMPLEX NUMBERS

defined so that $i^2 = -1$. It turns out that this will be enough to solve all polynomial equations, not just $x^2 + 1 = 0$.

To define the complex numbers let i to be a zero of the quadratic equation $x^2 + 1 = 0$, that is, $i = \sqrt{-1}$ is a square root of negative one. A **complex number** is any number that can be written in the **normal form** $a + bi$ where a and b are real numbers. The number a is called the **real part** of $a + bi$, and b is called the **imaginary part**. If b is zero, then $a + bi = a$ is real, so the real numbers are a subset of the complex numbers. If $a = 0$, then $a + bi = bi$ is called **pure imaginary** (or sometimes **purely imaginary**).

Two complex numbers $a + bi$ and $c + di$ are equal if and only if $a = c$ and $b = d$. The notions of "greater than" and "less than" do not apply to complex numbers, just to the real numbers. We denote the set of complex numbers by C.

We defined i by saying it is a zero of $x^2 + 1 = 0$. But this equation has two zeros—the other is $-i$. Because of this ambiguity we define the concept of "conjugation." The conjugate of i defined to be $-i$ and the conjugate of $-i$ to be i. Real numbers do not suffer from any ambiguity, so the conjugate of real number is itself. Putting these last statements together we get the following important definition: The **complex conjugate** of $a + bi$ is the number $a - bi$.

Complex conjugation is usually denoted by a horizontal bar:

$$\overline{a + bi} = a - bi.$$

For example, the conjugate of $\overline{2 - 3i}$ is $2 + 3i$, and a number z is real if and only $z = \overline{z}$ (see exercise 1a).

To add two complex numbers add the real and imaginary parts separately:

$$(a + bi) + (c + di) = (a + c) + (b + d)i.$$

For example, $(3 + 4i) + (2 + 5i) = 5 + 9i$, and $(1 + i) + (1 - i) = 2$.

To subtract two complex numbers subtract the real and imaginary parts separately:

$$(a + bi) - (c + di) = (a - c) + (b - d)i.$$

For example, $(3 + 4i) - (2 + 5i) = 1 - i$, and $(1 + i) - (1 - i) = 2i$.

To multiply two complex numbers multiply the numbers together as polynomials in i and replace i^2 by -1:

$$(a + bi)(c + di) = ac + (ad + bc)i + bdi^2 = (ac - bd) + (ad + bc)i.$$

For example, $(3 + 4i)(2 + 5i) = -14 + 23i$, and $(1 + i)(1 - i) = 2$.

2.1. Definition of the Complex Numbers

To divide two complex numbers multiply the both divisor and dividend by the conjugate of the divisor:

$$\frac{a+bi}{c+di} = \frac{a+bi}{c+di} \cdot \frac{c-di}{c-di} = \frac{(ac+bd)+(bc-ad)i}{c^2+d^2} = \frac{ac+bd}{c^2+d^2} + \frac{bc-ad}{c^2+d^2}i.$$

For example,

$$\frac{2+3i}{4-7i} = \frac{2+3i}{4-7i} \cdot \frac{4+7i}{4+7i} = \frac{-13+26i}{65} = -\frac{13}{65} + \frac{26}{65}i.$$

Example 2.1.1. *Express the following complex numbers in the normal form $a + bi$.*

1. $3 + (5 - 2i) + (12 - 23i) + (3 - i6)$
2. $(3 + 4i)(3 - 4i)$
3. $\dfrac{3i - 5}{2 + 7i}$
4. $(1 - i)^n$ *(for all positive integers n)*

Solution:

1. We add separately the real parts: $3, 5, 12, 3$; and the imaginary parts: $-2, -23, -6$ to get $23 - 31i$.

2. $(3 \cdot 3) + (3 \cdot -4i) + (4i \cdot 3) + (4i \cdot -4i) = 25$.

3. $\dfrac{3i - 5}{2 + 7i} = \dfrac{(-5 + 3i)(2 - 7i)}{(2 + 7i)(2 - 7i)} = \dfrac{11 + 41i}{53} = \dfrac{11}{53} + \dfrac{41}{53}i.$

4. First we will try small values of n hoping to see a pattern.

$$\begin{array}{ll}
(i-1)^1 = -1+i & (i-1)^5 = (-4)(-1+i) \\
(i-1)^2 = -2i & (i-1)^6 = (-4)(-2i) \\
(i-1)^3 = 2(1+i) & (i-1)^7 = (-4)2(1+i) \\
(i-1)^4 = -4 & (i-1)^8 = (-4)(-4)
\end{array}$$

We have been lucky! Since $(i-1)^4 = -4$, we know $(i-1)^{4m} = (-4)^m$ for all integers m. Now if we divide n by four, then we will get a quotient (call it m) and one of the four remainders: $0, 1, 2$ or 3. Using what we have found, we can now show that $(i-1)^n$ is as follows.

$$(i-1)^n = \begin{cases} (-4)^m & \text{if } n = 4m \\ (-1+i)(-4)^m & \text{if } n = 4m+1 \\ -2i(-4)^m & \text{if } n = 4m+2 \\ 2(1+i)(-4)^m & \text{if } n = 4m+3 \end{cases}$$

2. COMPLEX NUMBERS

Now that we can solve $x^2 + 1 = 0$, we can factor $A^2 + B^2$:

$$A^2 + B^2 = (A + Bi)(A - Bi).$$

In fact we will show in a later chapter that we can factor all polynomials completely using complex numbers. This result is called the Fundamental Theorem of Algebra.

The definition of the complex numbers was first published by the mathematician-physician Cardan in the sixteenth century. At that time, the properties of the negative integers were still being debated, hence the complex numbers were dubiously accepted and were dubbed **imaginary numbers**. Today the existence of imaginary numbers is in no more doubt than the existence of real numbers, but the old title remains.

Exercises 2.1:

1. Let z and w be complex numbers. Show each of the following.

   a) z is equal to its own conjugate if and only if it is real.

   b) z is equal to the negative of its own conjugate if and only if it is pure imaginary.

   c) The conjugate of the conjugate of z, $\bar{\bar{z}}$, is z.

   d) $z + \bar{z}$ is a real number.

   e) $z \cdot \bar{z}$ is a real number.

   f) $z - \bar{z}$ is a pure imaginary number.

   g) The product of conjugates is the conjugate of the product: $\bar{z} \cdot \bar{w} = \overline{z \cdot w}$.

   h) The quotient of conjugates is the conjugate of the quotient $\frac{\bar{z}}{\bar{w}} = \overline{\left(\frac{z}{w}\right)}$.

   i) The sum of conjugates is the conjugate of the sum: $\bar{z} + \bar{w} = \overline{z + w}$.

   j) The difference of conjugates is the conjugate of the difference: $\bar{z} - \bar{w} = \overline{z - w}$.

2. Express the following complex numbers in the normal form $a + bi$.

   a) $5 - i + (-3 + i) - (-5 - 4i)$
   b) $1/i$
   c) $(3 - 5i)(2 + 3i)$
   d) $(2 + i)/i$
   e) $(3 + i)(i - 3)/(i + i)$
   f) $(1 + i)/i + i/(1 - i)$
   g) $(3 - 4i)/(-i)$
   h) $(1 + i)^2/2$
   i) $(1 + i)^4/4$
   j) $(1 + i)^3/(1 - i)$
   k) $((-1 + \sqrt{3}i)/2)^2$
   l) $((-1 - \sqrt{3}i)/2)^3$
   m) $(\cos\theta + i\sin\theta)(\cos\theta - i\sin\theta)$ (this does not depend on θ).

3. For each of the following values of z and for all positive integers n, express z^n in the normal form $a + bi$.

   a) $z = i$
   b) $z = -i$
   c) $z = 1 + i$
   d) $z = 1 - i$
   e) $z = (1 + i)/\sqrt{2}$
   f) $z = (1 - i)/\sqrt{2}$
   g) $z = (-1 + \sqrt{3}i)/2$
   h) $z = (-1 - \sqrt{3}i)/2$

2.1. Definition of the Complex Numbers

4. Given that x and y are real numbers, solve the following for x and y.

    a) $2 + 3i - 4x + i(y - 2) = 4 - 5i$

    b) $(x + yi)(x - yi) = 0$

    c) $(2 - 3i)(x - yi) + (1 + i)(x + i) = 3ix - y$

    Hint: How do we tell if two complex numbers $a + bi$ and $c + di$ are equal?

5. (*) Show that if a, b, c, d and the product $(a+bi)(c+di)$ are all non-zero real numbers, then $c+di = k(a-bi)$ for some real constant k. Hint: Write the product in the normal form.

6. We say the real numbers $\mathbb{R}$ are ordered because they have the relation $>$ which satisfies the following three properties.

    a) For any $a \in \mathbb{R}$ exactly one of $a > 0$, $a = 0$, or $0 < a$ is true (called trichotomy).

    b) If $a > 0$ and $b > 0$, then $a + b > 0$ and $a \cdot b > 0$.

    c) If $a > b$, then $a + c > b + c$ for any $c \in \mathbb{R}$.

    Show that $\mathbb{C}$ cannot be ordered. Hint: If $z < 0$, then by adding $-z$ to each side we see $0 < -z$. Use this to show both $z^2 > 0$ (if $z \neq 0$) and $1 > 0$. What happens if $z = i$?

2. Complex Numbers

2.2 The Complex Plane

The real number line serves as a model for the real numbers. That is, we can think of the real numbers as points on a line. Similarly complex numbers can be represented by a plane called the plane of complex numbers, or just the **complex plane**. We construct the complex plane by associating to each complex number $a + bi$ the point with cartesian coordinate (a, b). Thus 0 corresponds to the origin and i corresponds to $(0, 1)$ (Figure 2.2). Because of this correspondence the normal form of a complex

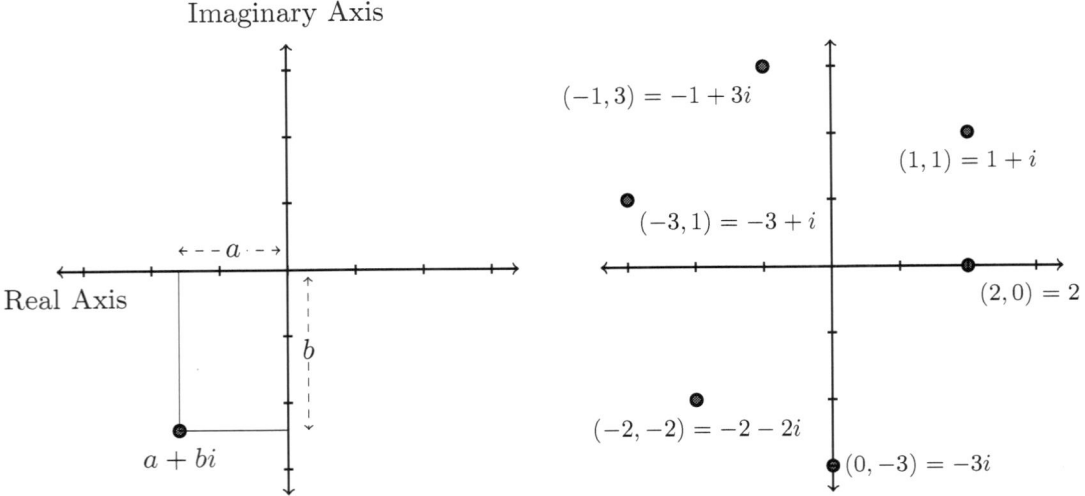

Figure 2.2: The Complex Plane $\mathbb{C}$ (and Example Points)

number $a + bi$ is also called the **rectangular form**.

On the complex plane the horizontal axis (x-axis) is called the real axis, and the vertical axis (y-axis) is called the imaginary axis. Notice that the real axis is also the real number line.

To give the operation of addition a geometric meaning we associate the complex number $z = a + bi$ with the vector pointing from the origin to (a, b). Then the addition of complex numbers may be interpreted as vector addition (see Figure 2.3). In the same way we may think of $u - z$ as vector subtraction.

The **absolute value** (or **modulus**) of the complex number z is the modulus of the associated vector. That is, the distance between z and the origin. Using the Pythagorean Theorem we see that this is $|a + bi| = \sqrt{a^2 + b^2}$. It is useful to notice that
$$|a + bi|^2 = a^2 + b^2 = (a + bi)(a - bi), \text{ that is, } |z|^2 = z\bar{z}.$$

Warning: The notions of "greater than" and "less than" may only be applied to real numbers. However the modulus of a complex number is a real number, so we can speak of one complex number having a greater magnitude (modulus) than another.

2.2. The Complex Plane

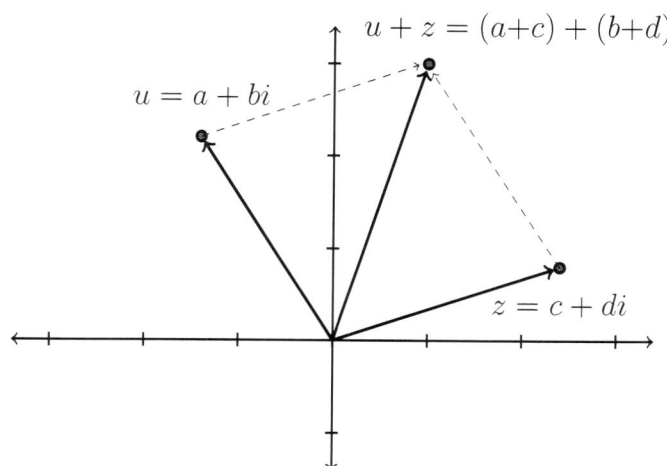

Figure 2.3: Complex Numbers–Addition

Exercises 2.2:

1. Graph each of the following pair of numbers and their sum as vectors in the complex plane.

   a) $1+2i$, $4-2i$
   b) 2, $2i$
   c) $1+\sqrt{3}i$, $1-\sqrt{3}i$
   d) $1+i$, $1-i$
   e) 3, $-2i$
   f) $1+2i$, $-1-2i$
   g) $\sqrt{3}+i$, $\sqrt{3}-i$
   h) $1+i$, $1+i$
   i) $3.2-2.1i$, $0.8+0.6i$

2. Find the absolute value of each of the following.

   a) $23 - 45i$
   b) $3 + 4i$
   c) $(\sqrt{3} - i)/2$
   d) $16i$
   e) $\cos\theta + i\sin\theta$
   f) $(a+bi)/(a-bi)$
   g) $(1+i)^6$
   h) $(1+i)/\sqrt{2}$
   i) $\tan\theta + i$

3. Show that the product of the absolute value of two complex numbers is the absolute value of their products, that is,
$$|z_1| \cdot |z_2| = |z_1 \cdot z_2|.$$

4. (*) Extend the result of the previous problem to the product of n complex numbers.
$$|z_1| \cdot |z_2| \cdot |z_3| \cdot \ldots \cdot |z_n| = |z_1 \cdot z_2 \cdot z_3 \cdot \ldots \cdot z_n|.$$

   Hint: Use induction.

5. Show that the quotient of the absolute values of two complex numbers is the absolute value of their quotients, that is if $z_2 \neq 0$, then
$$\frac{|z_1|}{|z_2|} = \left|\frac{z_1}{z_2}\right|.$$

2. COMPLEX NUMBERS

6. Prove the following fundamental theorem.

   Theorem 2.2.1 (Triangle Inequality). *Let z_1 and z_2 be any complex numbers. Then*
   $$|z_1 + z_2| \leq |z_1| + |z_2|.$$

   Hint: Use the correspondence between the addition of vectors and the addition of complex numbers. Where did this inequality get its name?

7. Prove the following corollary of the Triangle Inequality.

   Corollary 2.2.2. *Let z_1, z_2 be any complex numbers. Then*
   $$|z_1| - |z_2| \leq |z_1 - z_2|.$$

   Hint: Write $z_1 = z_2 + (z_1 - z_2)$, and use the triangle inequality.

8. Extend the triangle inequality by proving the following.

   Corollary 2.2.3. *Let $z_1, z_2, \ldots, z_n$ be any complex numbers. Then*
   $$|z_1 + z_2 + \ldots + z_n| \leq |z_1| + |z_2| + \ldots + |z_n|.$$

   Hint: Use induction.

9. Give a geometric interpretation of the following. Note that $\Re(z)$ is a symbol for the real part of z and $\Im(z)$ is a symbol for the imaginary part of z. (If you find these too difficult to write, many mathematicians use $\text{Re}(z)$ and $\text{Im}(z)$ respectively.)

   a) The relationship between $a + bi$ and $a - bi$.
   b) The relationship between z and $|z|$.
   c) The relationship between z and $-z$.
   d) The numbers for which $|z| < 1$.
   e) The numbers for which $|z| > 1$.
   f) The numbers for which $|z| = 1$.
   g) The numbers for which $\Re(z) = 2$.
   h) The numbers for which $\Re(z) > 2$.
   i) The numbers for which $|\Re(z)| \leq 2$.
   j) The numbers for which $\Im(z) = -3$.
   k) The numbers for which $\Im(z) > -3$.
   l) The numbers for which $|\Im(z)| \leq -3$.

2.2. The Complex Plane

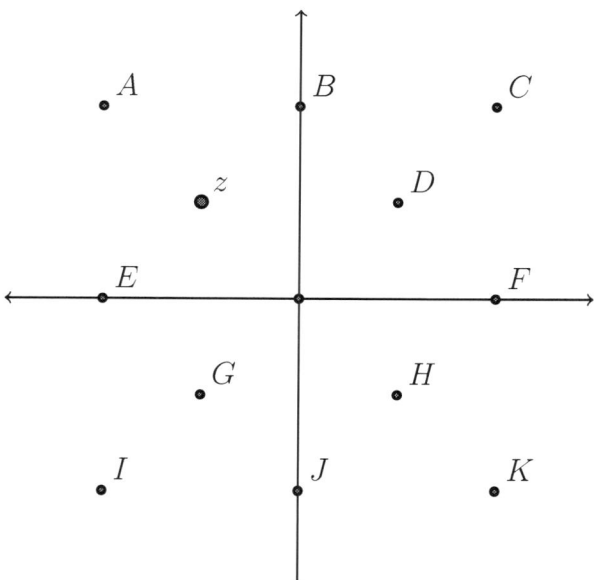

Figure 2.4: Graph for Problem 10

10. Let z be as indicated in Figure 2.4. Of the numbers $A, B, \ldots, K$ which one is

    a) $\bar{z}$
    b) $-\bar{z}$
    c) $2\bar{z}$
    d) $-2\bar{z}$
    e) $\bar{z} + z$
    f) $\bar{z} - z$
    g) $-z - \bar{z}$
    h) $\bar{z} - z$

11. (*) Explain the error in the following false "proof" that $-1 = 1$.

    $$-1 = i \cdot i = (-1)^{1/2}(-1)^{1/2} = ((-1)(-1))^{1/2} = 1^{1/2} = 1.$$

    To make it easier for you to write your response, below we marked the different equalities with letters—just explain which are in error.

    $$-1 \stackrel{a}{=} i \cdot i \stackrel{b}{=} (-1)^{1/2}(-1)^{1/2} \stackrel{c}{=} ((-1)(-1))^{1/2} \stackrel{d}{=} 1^{1/2} \stackrel{e}{=} 1.$$

12. (*) Explain the error in the following false "proof" that $-1 = 1$.

    $$-1 = (-1)^1 = (-1)^{2 \cdot \frac{1}{2}} = ((-1)^2)^{\frac{1}{2}} = 1^{\frac{1}{2}} = 1.$$

    As in the previous problem, to make it easier for you to write your response, we mark the different equalities.

    $$-1 \stackrel{a}{=} (-1)^1 \stackrel{b}{=} (-1)^{2 \cdot \frac{1}{2}} \stackrel{c}{=} ((-1)^2)^{\frac{1}{2}} \stackrel{d}{=} 1^{\frac{1}{2}} \stackrel{e}{=} 1.$$

2. Complex Numbers

2.3 The Polar Form of Complex Numbers

Each point $z = a + bi$ on the complex plane may also be associated with its polar coordinates (r, θ), where $r = |z| = \sqrt{a^2 + b^2}$, $a = r\cos\theta$, and $b = r\sin\theta$. So we may write $z = a + bi$ in the **polar form**:

$$z = r(\cos\theta + i\sin\theta).$$

When z is written in the polar form, r is called the **modulus** of z, and θ is called an

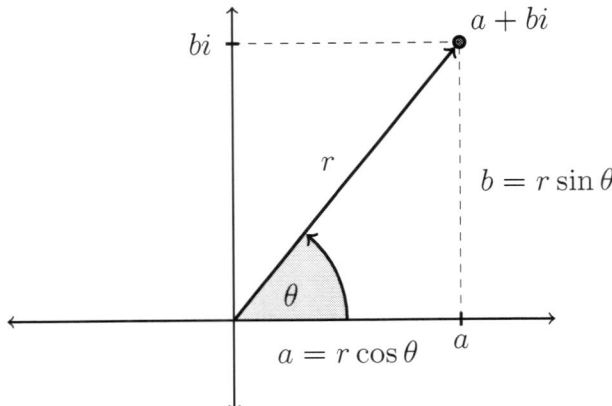

Figure 2.5: Complex Numbers–Polar Form

argument (or **amplitude**) of z (see Figure 2.5). It is useful to note that $\tan\theta = \dfrac{b}{a}$, but because of the way the inverse tangent is defined, we do not always have that $\theta = \tan^{-1}\left(\dfrac{b}{a}\right)$.

Example 2.3.1. *Convert the following complex numbers to polar form: 1) $1 + i$, 2) $-4i$ and 3) $3 - 4i$.*

Solution: We first find the modulus r, and then an angle θ that satisfies *both* $\cos\theta = a/r$ and $\sin\theta = b/r$. Do not be led astray by just checking one of them!

1. First $r = |1+i| = \sqrt{2}$, so $\cos\theta = \sin\theta = 1/\sqrt{2}$, or $\theta = \pi/4 + 2n\pi$ for any integer n. Each choice of n gives a valid representation of the same point, so why not choose the simplest: $n = 0$? With $n = 0$ we have

$$1 + i = \sqrt{2}\left(\cos\frac{\pi}{4} + i\sin\frac{\pi}{4}\right).$$

   This shows one representation in polar form is $(r, \theta) = (\sqrt{2}, \pi/4)$.

2.3. The Polar Form of Complex Numbers

2. Note $r = |4i| = 4$, so $\cos\theta = 0$ and $\sin\theta = -1$. Thus $\theta = -\pi/2 + 2n\pi$ for any integer n. Again we choose $n = 0$, so
$$4i = 4\left(\cos\left(-\frac{\pi}{2}\right) + i\sin\left(-\frac{\pi}{2}\right)\right).$$
This gives us the polar form $(r,\theta) = (4, -\pi/2)$. Instead we could have chosen $n = 1$ so that the resulting angle, $\theta = 3\pi/2$, would have been positive.

3. Finally $r = \sqrt{3^2 + 4^2} = 5$, so $\cos\theta = 3/5$, $\sin\theta = -4/5$, and $\tan\theta = -4/3$. One possible value of θ is approximately 2.4981 radians. So
$$3 - 4i \approx 5(\cos 2.4981 + i\sin 2.4981),$$
or $(r,\theta) \approx (5, 2.4981)$.

∎

Example 2.3.2. *Convert the following to normal form:*

1. $2\left(\cos\dfrac{-11\pi}{6} + i\sin\dfrac{-11\pi}{6}\right)$

2. $\dfrac{2}{3}(\cos 60° + i\sin 60°)$.

Solution: We simply find $a = r\cos\theta$ and $b = r\sin\theta$.

1. Note $a = 2\cos\dfrac{-11\pi}{6} = 2\dfrac{\sqrt{3}}{2} = \sqrt{3}$ and $b = 2\sin\dfrac{-11\pi}{6} = 1$. So the normal form is $\sqrt{3} + i$.

2. For this example $a = \dfrac{2}{3}\cos 60° = \dfrac{1}{3}$ and $b = \dfrac{2}{3}\sin 60° = \dfrac{1}{\sqrt{3}}$, so the desired number is $2 + \dfrac{1}{\sqrt{3}}i$.

∎

Defining arguments

The trigonometric functions are periodic so the polar form of a complex number is not unique; in fact
$$(r,\theta) = (r, \theta + 2n\pi)$$
for all integers n. In other words, if θ is an argument of z, then so is $\theta + 2n\pi$ for all integers n. Because of this, "the argument" of a complex number is not well defined:

17

2. COMPLEX NUMBERS

0, 2π, -10π and 2000π are all arguments of the number 1. We define *the set of arguments* of the number z as follows.

$$\arg z = \{\,\theta \in \mathbb{R} \mid z = r(\cos\theta + i\sin\theta) \text{ for } r > 0\,\}$$

Still, sometimes it is helpful to select a single element of $\arg(z)$ to be **the principle argument of** z, denoted $\operatorname{Arg} z$. The standard choice is to let the $\operatorname{Arg} z$ be the argument of $z \neq 0$ for which $-\pi < \operatorname{Arg} z \leq \pi$.

Example 2.3.3. *Find* $\arg(-1+i)$ *and* $\operatorname{Arg}(-1+i)$.

Solution: It is true that if θ is an argument for $-1+i$, then $\tan\theta = -1/1$. However, $-1+i$ is in the second quadrant (see Figure 2.6), so $\theta \neq -\pi/4$. Instead we must add π to get $\operatorname{Arg}(-1+i) = 3\pi/4$. From this we have

$$\arg(-1+i) = \left\{\,\frac{3\pi}{4} + 2\pi n \,\middle|\, n \in \mathbb{Z}\,\right\}.$$

■

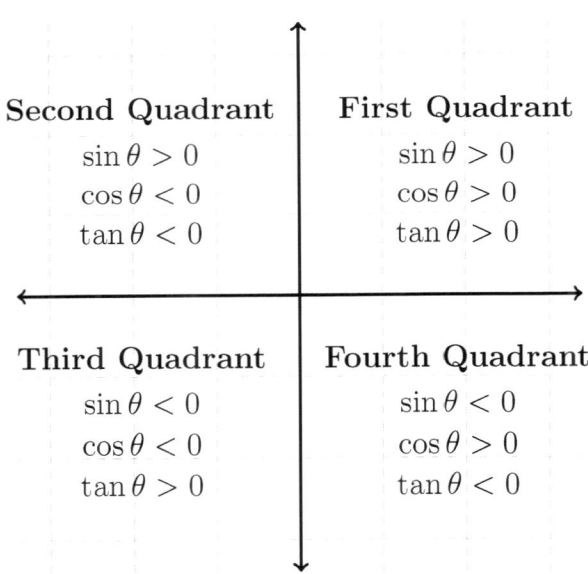

Figure 2.6: The Four Quadrants of the Plane ($\mathbb{C}$ or $\mathbb{R}^2$)

Arithmetic Operations

Using the angle addition formulas from trigonometry we may show how to multiply and divide complex numbers in polar form (see exercise 1). To add or subtract

2.3. The Polar Form of Complex Numbers

numbers in polar forms is rather difficult. It is usually easier to just express them in norm form and then add or subtract as desired.

To Multiply Complex Numbers in Polar Form: The modulus of the product is the product of the moduli, and an argument of a product is the sum of the arguments. That is,

$$r_1(\cos\theta_1 + i\sin\theta_1) \cdot r_2(\cos\theta_2 + i\sin\theta_2) = r_1 r_2 \left(\cos(\theta_1 + \theta_2) + i\sin(\theta_1 + \theta_2)\right).$$

To Divide Complex Numbers in Polar Form: The modulus of the quotient is the quotient of the moduli, and an argument of the quotient is the difference of the arguments. That is,

$$\frac{r_1(\cos\theta_1 + i\sin\theta_1)}{r_2(\cos\theta_2 + i\sin\theta_2)} = \frac{r_1}{r_2}\left(\cos(\theta_1 - \theta_2) + i\sin(\theta_1 - \theta_2)\right).$$

Exercises 2.3:

1. Prove the rule for the multiplication of complex numbers in polar form.

2. Prove the rule for the division of complex numbers in polar form.

3. Prove that for all non-zero complex numbers that (a) $\arg(z \cdot w) = \arg(z) + \arg(w)$ and (b) $\arg(z/w) = \arg(z) - \arg(w)$.

4. Express the following numbers in polar form (r, θ).

   a) -5
   b) i
   c) $-i$
   d) $-1 + i$
   e) $-2 - 2i$
   f) $3i + 3$
   g) $-1 + \sqrt{3}i$
   h) $-1 - \sqrt{3}i$
   i) $\sqrt{3} - i$
   j) $4 + 3i$
   k) $4 - i$
   l) $-1 + 4i$

5. Each of the following pairs r, θ represent the polar coordinates of a complex number. Express each number in the normal form $a + bi$. *Do not approximate!*

   a) $3, 0$
   b) $2, \pi/12$
   c) $4, \pi/6$
   d) $2, 45°$
   e) $2, 60°$
   f) $8, 75°$
   g) $1, 90°$
   h) $10, 7\pi/12$
   i) $2, 2\pi/3$
   j) $4, 3\pi/4$
   k) $6, 5\pi/6$
   l) $12, 11\pi/12$
   m) $345, -13\pi$
   n) $6, 4\pi/3$
   o) $2, 270°$
   p) $200, 5\pi/4$
   q) $215, 7\pi/4$
   r) $5, 62\pi$

   Hint: $\cos\dfrac{\pi}{12} = \dfrac{\sqrt{6} + \sqrt{2}}{4}$, $\sin\dfrac{\pi}{12} = \dfrac{\sqrt{6} - \sqrt{2}}{4}$.

6. Each of the following pairs r, θ represent the polar coordinates of a complex number. Express each number in the normal form $a + bi$. Approximate a and b, expressing the results correct to six decimal places. (Note that the last six are in radians.)

2. COMPLEX NUMBERS

a) $5, 10°$
b) $36, 12°$
c) $8, 96°$
d) $2, 72°$
e) $12, 36°$
f) $45, 20°$
g) $7, 200°$
h) $10, 400°$
i) $55, 293°$
j) $7, 20$
k) $10, 400$
l) $55, 0.25$
k) $18, 0.213$
l) $3, 141$
m) $15, 0.002$

7. In which quadrants is $\tan^{-1}(b/a)$ an argument of $a + bi$?

8. Show that the principle argument for $z = a + bi$ can be found as follows.

$$\operatorname{Arg} z = \begin{cases} \tan^{-1}(y/x) & \text{if } x > 0, \\ \tan^{-1}(y/x) + \pi & \text{if } x < 0 \text{ and } y \geq 0, \\ \tan^{-1}(y/x) - \pi & \text{if } x < 0 \text{ and } y < 0, \\ \pi/2 & \text{if } x = 0 \text{ and } y > 0, \\ -\pi/2 & \text{if } x = 0 \text{ and } y < 0, \text{ and} \\ \text{indeterminant} & \text{if } x = 0 \text{ and } y = 0. \end{cases}$$

9. Show that if θ is an argument of z, then

   a) $\pi + \theta$ is an argument of $-z$.

   b) $-\theta$ is an argument of $1/z$.

   c) 2θ is an argument of z^2.

   d) $n\theta$ is an argument of z^n.

10. Recall that we can make the argument of the complex number $z \neq 0$ well defined by letting $\operatorname{Arg} z$ be the principle argument of z, that is, the argument of z which is in $(-\pi, \pi]$.

    a) Give an example of complex numbers z and w for which $\operatorname{Arg}(z \cdot w) = \operatorname{Arg}(z) + \operatorname{Arg}(w)$ and another for which $\operatorname{Arg}(z \cdot w) \neq \operatorname{Arg}(z) + \operatorname{Arg}(w)$.

    b) Give an example of complex numbers z and w for which $\operatorname{Arg}(z/w) = \operatorname{Arg}(z) - \operatorname{Arg}(w)$ and another for which $\operatorname{Arg}(z/w) \neq \operatorname{Arg}(z) - \operatorname{Arg}(w)$.

    c) Shade the regions of the complex plane for $\operatorname{Arg}(z^2) = 2\operatorname{Arg}(z)$, that is, $\operatorname{Arg}(z \cdot z) = \operatorname{Arg}(z) + \operatorname{Arg}(z)$. Mark the boundaries with dashed or solid curves as appropriate.

2.4 Euler's Formula

Recall that the functions e^x, $\cos x$ and $\sin x$ may each be written as infinite series:

$$e^x = 1 + x + \frac{x^2}{2} + \frac{x^3}{3!} + \frac{x^4}{4!} + \frac{x^5}{5!} + \cdots$$

$$\cos x = 1 - \frac{x^2}{2} + \frac{x^4}{4!} - \frac{x^6}{6!} + \frac{x^8}{8!} + \cdots$$

$$\sin x = x - \frac{x^3}{3!} + \frac{x^5}{5!} - \frac{x^7}{7!} + \frac{x^9}{9!} + \cdots$$

Notice the remarkable similarity between these three expansions. If it was not for the signs, then we could add the last pair of series to get the first. We can remove the sign problem by replacing x with iy. This gives the following fundamental theorem (see exercise 1):

Theorem 2.4.1 (Euler's Formula). $e^{x+iy} = e^x(\cos y + i \sin y)$

(The name Euler is pronounced "oiler.")

Example 2.4.1. *Express the following in normal form.*

1) $e^{\pi i}$ 2) $e^{\pi i/2}$ 3) $e^{\pi i/3}$ 4) $e^{3+\pi i/4}$

Solution: Using Euler's Formula we find the following.

1. $e^{\pi i} = \cos \pi + i \sin \pi = -1$
2. $e^{\pi i/2} = \cos \frac{\pi}{2} + i \sin \frac{\pi}{3} = i$
3. $e^{\pi i/3} = \cos \frac{\pi}{3} + i \sin \frac{\pi}{3} = \frac{1}{2} + \frac{\sqrt{3}}{2} i$
4. $e^{3+\pi i/4} = e^3 \left(\cos \frac{\pi}{4} + i \sin \frac{\pi}{4} \right) = \frac{e^3}{\sqrt{2}}(1+i) = \frac{e^3}{\sqrt{2}} + \frac{e^3}{\sqrt{2}} i$

■

The equation $e^{\pi i} + 1 = 0$ that we found in part (1) of this example is perhaps the most beautiful equation in all of mathematics, relating the additive identity 0, the multiplicative identity 1, the ratio of the circumference to the diameter of a circle π, the complex number i and the base of the natural logarithms e.

Using Euler's Formula, we notice the magnitude of e^z is determined by just the real part of $z = x + iy$ since $|e^z| = e^x$. Similarly, the argument of e^z is determined

2. COMPLEX NUMBERS

just by the imaginary part of z. In fact, one argument of e^z is y. An important consequence of these facts is that the points on the unit circle are given by

$$e^{i\alpha} = \cos\alpha + i\sin\alpha,$$

which is associated with the point $(\cos\alpha, \sin\alpha)$ in rectangular coordinates. A second consequence is that the complex number z with polar coordinates (r, α) may be written $re^{i\alpha}$. This form, $re^{i\alpha}$, is also called the polar form of z (see Figure 2.7).

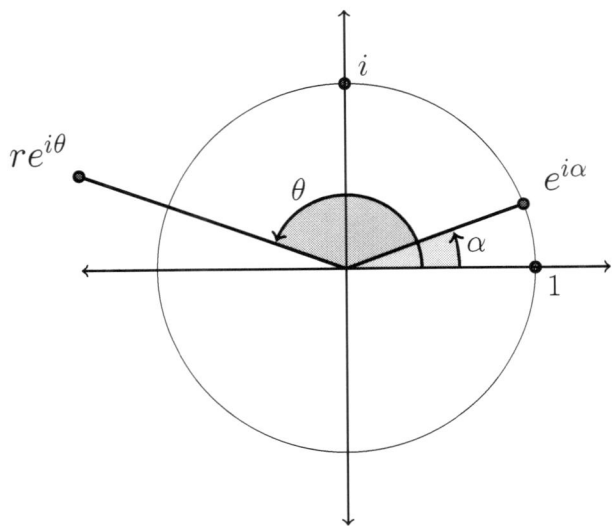

Figure 2.7: Complex Numbers–Exponential Form

Example 2.4.2. *Express the following in polar form:*

1) $1 - \sqrt{3}i$ 2) -45 3) $i/7$ 4) $3 + 4i$

Solution: We wish to write these as $re^{i\alpha}$. Here r is just the modulus of the number $z = a + bi$, which is very easy to calculate. Also α is an argument, which can be found using any two of the following.

$$\sin\alpha = \frac{b}{r}, \qquad \cos\alpha = \frac{a}{r}, \qquad \tan\alpha = \frac{b}{a}.$$

Since we know which quadrant z is in, we can use just one of these relations—being careful to pick an angle α lying in that quadrant.

1. Note $r = \sqrt{1^2 + \sqrt{3}^2} = 2$ and $\tan\alpha = -\sqrt{3}$, so α is $\frac{\pi}{3} + n\pi$ for some integer n. The number is in the fourth quadrant so we may choose $\alpha = -\frac{\pi}{3}$.

$$1 - \sqrt{3}i = 2e^{-\pi i/3}$$

2. Clearly $r = 45$. For α we may choose any odd multiple of π, so

$$-45 = 45e^{\pi i}.$$

In general, for numbers like -45 or $i/7$, it is easier to picture where these numbers lie on the plane, than to use the trigonometric relations.

3. Clearly $\frac{i}{7} = \frac{1}{7}e^{\pi i/4}$.

4. We easily find $r = 5$. It is also easy to picture where $3 + 4i$ is on the complex plane, but its argument(s) are not among our standard angles. However, the number is in the first quadrant, so if we define $\alpha = \tan^{-1}\left(\frac{4}{3}\right)$, then $3 + 4i = 5e^{i\alpha}$. Depending on our application, we would decide whether or not to approximate $\tan^{-1}\left(\frac{4}{3}\right)$, thereby approximate α.

■

Another notation for $e^{i\alpha}$ is **cis α**, this comes from the c, the i and the s in $\cos\alpha + i\sin\alpha$. This "cis" notation is common in older elementary textbooks, but is rare in mathematical journals.

Exercises 2.4:

1. Prove Euler's Formula. Hint: Replace the x in the three infinite series with iy to show that

$$e^{iy} = \cos y + i\sin y.$$

Extend this result to cover e^{x+iy}.

2. For each of the following arguments α, express $e^{i\alpha}$ in the form $a + bi$ (do not approximate).

   a) 0 b) $\pi/12$ c) $\pi/6$
   d) $\pi/4$ e) $\pi/3$ f) $5\pi/12$
   g) $\pi/2$ h) $7\pi/12$ i) $2\pi/3$
   j) $3\pi/4$ k) $5\pi/6$ l) $11\pi/12$
   m) π n) $4\pi/3$ o) $3\pi/2$
   p) $5\pi/4$ q) $7\pi/4$ r) 62π

   Hint $\cos\dfrac{\pi}{12} = \dfrac{\sqrt{6}+\sqrt{2}}{4}$, $\sin\dfrac{\pi}{12} = \dfrac{\sqrt{6}-\sqrt{2}}{4}$.

3. For each of the following arguments α, express $e^{i\alpha}$ in the form $a + bi$ (you may approximate).

2. COMPLEX NUMBERS

a) 10°   b) 12°   c) 96°
d) 72°   e) 36°   f) 20°
g) 200°   h) 400°   i) 293°

4. Let $z = re^{i\alpha}$ with $r \geq 0$. Express in the same form: a) $-z$, b) $|z|$ and c) $\bar{z}$.

5. Let $z = r \operatorname{cis} \alpha$ with $r \geq 0$. Express in the same form: a) $-z$, b) $|z|$ and c) $\bar{z}$.

6. Let $z = re^{i\alpha}$ with $r \geq 0$. Express in the same form: a) z^2, b) z^3, c) z^4, d) z^n for any positive integer n.

7. Let $z = r \operatorname{cis} \alpha$ with $r \geq 0$. Express in the same form: a) z^2, b) z^3, c) z^4, d) z^n for any positive integer n.

8. (*) Recall that in calculus we defined $\cosh z = (e^z + e^{-z})/2$, $\sinh z = (e^z - e^{-z})/2$. Let $a = a + bi$ and prove the following identities:

   a) $\sin z = \frac{1}{2i}(e^{iz} - e^{-iz})$
   b) $\cos z = \frac{1}{2}(e^{iz} + e^{-iz})$
   c) $\sin z = -i \sinh iz$
   d) $\cos z = \cosh iz$
   e) $\sin z = \sin a \cosh b + i \cos a \sinh b$
   f) $\cos z = \cos a \cosh b + i \sin a \sinh b$

   Hint: Recall $e^{iz} = \cos z + i \sin z$, so $e^{-iz} = \cos z - i \sin z$. Use these together to solve for $\sin z$ and $\cos z$.

2.5 De Moivre's Theorem

A simple, but very important, consequence of Euler's Formula is the following.

Theorem 2.5.1 (De Moivre's Theorem). *Let n be an integer. Then*

$$\cos n\alpha + i \sin n\alpha = (\cos \alpha + i \sin \alpha)^n$$

Proof. By Euler's Formula twice we have

$$\cos n\alpha + i \sin n\alpha = e^{n\alpha i} = \left(e^{\alpha i}\right)^n = (\cos \alpha + i \sin \alpha)^n.$$

Almost too easy! □

2.5. De Moivre's Theorem

Example 2.5.1. *Find $\cos 5\alpha$ in terms of $\cos \alpha$ and $\sin \alpha$.*

Solution: By De Moivre's Theorem we have
$$\cos 5\alpha + i \sin 5\alpha = (\cos \alpha + i \sin \alpha)^5$$
which is equal to
$$\cos^5 \alpha + 5i \cos^4 \alpha \sin \alpha - 10 \cos^3 \alpha \sin^2 \alpha - 10i \cos^2 \alpha \sin^3 \alpha + 5 \cos \alpha \sin^4 \alpha + i \sin^5 \alpha$$
Comparing the real and imaginary parts we see
$$\cos 5\alpha = \cos^5 \alpha - 10 \cos^3 \alpha \sin^2 \alpha + 5 \cos \alpha \sin^4 \alpha,$$
and
$$\sin 5\alpha = 5 \cos^4 \alpha \sin \alpha - 10 \cos^2 \alpha \sin^3 \alpha + \sin^5 \alpha.$$

■

Now we can find all the multiple angle formulas easily!

Example 2.5.2. *Find $\cos(2\theta - \alpha)$ in terms of $\cos \theta$, $\cos \alpha$, $\sin \theta$, and $\sin \alpha$.*

Solution: By Euler's Formula:
$$\begin{aligned} e^{i(2\theta - \alpha)} &= (e^{i\theta})^2 e^{-i\alpha} \\ &= (\cos \theta + i \sin \theta)^2 (\cos \alpha - i \sin \alpha) \\ &= \cos^2 \theta \cos \alpha + 2i \cos \theta \sin \theta \cos \alpha - \sin^2 \theta \cos \alpha + \\ & \quad -i \cos^2 \theta \sin \alpha - 2 \cos \theta \sin \theta \sin \alpha - i \sin^2 \theta \sin \alpha. \end{aligned}$$
The real part of this expression is $\cos(2\theta - \alpha)$; that is,
$$\cos(2\theta - \alpha) = \cos^2 \theta \cos \alpha - \sin^2 \theta \cos \alpha - 2 \cos \theta \sin \theta \sin \alpha.$$

■

Example 2.5.3. *Find $(1 + i)^{312}$.*

Solution: It is easy to see that $1 + i = 2^{1/2}\left(\cos \frac{\pi}{4} + i \sin \frac{\pi}{4}\right)$. By De Moivre's Theorem,
$$\begin{aligned} (1 + i)^{312} &= (2^{1/2})^{312} \left(\cos\left(312 \left(\frac{\pi}{4}\right)\right) + i \sin\left(312 \left(\frac{\pi}{4}\right)\right)\right) \\ &= 2^{156} (\cos 78\pi + i \sin 78\pi) = 2^{156}. \end{aligned}$$

■

2. COMPLEX NUMBERS

Exercises 2.5:

1. Express the following numbers in the form $re^{i\theta}$ with $r \geq 0$. In parts (a) through (i) and (l) do not approximate.

   a) -5
   b) i
   c) $-i$
   d) $-1 + i$
   e) $-2 - 2i$
   f) $3i + 3$
   g) $-1 + \sqrt{3}i$
   h) $-1 + \sqrt{3}i$
   i) $\sqrt{3} - i$
   j) $4 + 3i$
   k) $4 - i$
   l**) $(\sqrt{3}+1) + (\sqrt{3}-1)i$

2. Use De Moivre's Theorem or Euler's Formula to find each of the following in terms of $\sin \beta$ and $\cos \beta$.

   a) $\cos 2\beta$
   b) $\sin 2\beta$
   c) $\cos 4\beta$
   d) $\sin 4\beta$
   e) $\cos 5\beta$
   f) $\sin 5\beta$
   g) $\cos 7\beta$
   h) $\sin 7\beta$
   i) $\tan 4\beta$

3. Use De Moivre's Theorem and Euler's Formula to express each of the following in terms of $\sin \beta$, $\sin \theta$, $\cos \beta$ and $\cos \theta$.

   a) $\sin(\beta + \theta)$
   b) $\cos(\beta + \theta)$
   c) $\tan(\beta + \theta)$
   d) $\sin(2\beta + \theta)$
   e) $\cos(2\beta + \theta)$
   f) $\tan(2\beta + \theta)$
   g) $\sin(2\beta - 3\theta)$
   h) $\cos(2\beta - 3\theta)$
   i) $\tan(2\beta - 3\theta)$
   j) $\sin 3\theta$
   k) $\cos 4\theta$
   l) $\sin 5\theta$

4. For each of the following pairs z, n, express z^n *exactly* in the normal form $a + bi$.

   a) $1 + i$; 23
   b) $2^{-1/2}(1 + i)$; -234
   c) $2^{-1/2}(1 + i)$; 330
   d) $3^{1/2} - i$; 3
   e) $(3^{1/2} - i)/2$; 26
   f) $2 \operatorname{cis} 15°$; 6
   g) $2 \operatorname{cis} 7°$; 360
   h) $\cos \frac{\pi}{7} + i \sin \frac{\pi}{7}$; 14

5. For each of the following pairs z, n, express z^n in the normal form $a + bi$. You may approximate.

   a) $2 + 3i$; 7
   b) $1 + 2i$; 6
   c) $5 - 2i$; 4
   d) $1 - 2i$; 10
   e) $3 \operatorname{cis}(27°)$; 5
   f) $\operatorname{cis} 1°$; 413

2.6 Roots of Complex Numbers

Using De Moivre's Theorem we may find all the nth roots of a complex number z, that is, all the solutions to

$$\alpha^n = z.$$

To do this let (R, θ) be the polar coordinates of z, so

$$\alpha^n = z = Re^{(\theta + 2m\pi)i}$$

2.6. Roots of Complex Numbers

for every integer m. Taking the nth root we find

$$\alpha = R^{1/n} e^{(\theta+2m\pi)i/n} = R^{1/n}\left(\cos\left(\frac{\theta+2m\pi}{n}\right) + i\sin\left(\frac{\theta+2m\pi}{n}\right)\right)$$

for every integer m. Only n of these roots are distinct, so we usually choose those with $m = 0, 1, 2, \ldots, n-1$.

Example 2.6.1. *Find and graph the cube roots of i.*

Solution: $i = e^{(4m+1)\pi i/2}$ so the cube roots of i are $e^{(4m+1)\pi i/6}$ ($m = 0, 1, 2$). So

$$\cos\frac{\pi}{6} + i\sin\frac{\pi}{6} = \frac{\sqrt{3}}{2} + \frac{1}{2}i$$

$$\cos\frac{5\pi}{6} + i\sin\frac{5\pi}{6} = -\frac{\sqrt{3}}{2} + \frac{1}{2}i$$

$$\cos\frac{9\pi}{6} + i\sin\frac{9\pi}{6} = -i.$$

We graph these on the complex plane in Figure 2.8. ∎

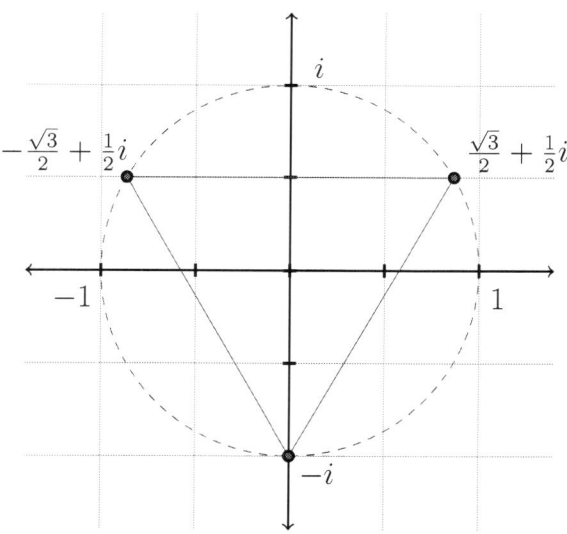

Figure 2.8: The Three Cube-Roots of i

2. COMPLEX NUMBERS

In the special case that $z = 1$, the nth roots of z are called the nth **roots of unity**.

Example 2.6.2. *Find and graph the sixth roots of unity.*

Solution: The nth roots of unity are given by:

$$\cos\left(\frac{\theta + 2m\pi}{n}\right) + i\sin\left(\frac{\theta + 2m\pi}{n}\right) \text{ for } m = 0, 1, 2, ..., n-1.$$

So the sixth roots of unity are

$$\cos(0) + i\sin(0) = 1$$
$$\cos(\pi/3) + i\sin(\pi/3) = \frac{1}{2} + \frac{\sqrt{3}}{2}i$$
$$\cos(2\pi/3) + i\sin(2\pi/3) = -\frac{1}{2} + \frac{\sqrt{3}}{2}i$$
$$\cos(\pi) + i\sin(\pi) = -1$$
$$\cos(4\pi/3) + i\sin(4\pi/3) = -\frac{1}{2} - \frac{\sqrt{3}}{2}i$$
$$\cos(5\pi/3) + i\sin(5\pi/3) = \frac{1}{2} - \frac{\sqrt{3}}{2}i.$$

We graph these on the complex plane in Figure 2.9. ∎

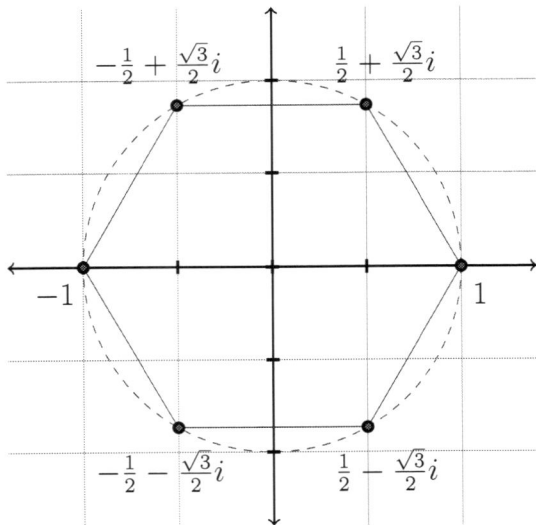

Figure 2.9: The Six Sixth-Roots of Unity

2.6. Roots of Complex Numbers

Notice that in Example 2.6.1 the cube roots of i are the vertices of an equilateral triangle (Figure 2.8), and that in in Example 2.6.2 the sixth roots of unity are the vertices of a hexagon (Figure 2.9). The following theorem shows that this is true in general.

Theorem 2.6.1. *Let z be a non-zero complex number and n be an integer greater than two. The nth roots of z are the vertices of a regular polygon with n sides inscribed in the circle centered on the origin and with radius $|z|^{1/n}$.*

Exercises 2.6:

1. Find exactly and then graph the nth roots of unity for (a) $n = 2$, (b) $n = 3$, (c) $n = 4$, (d*) $n = 5$, and (e*) $n = 8$.

2. Approximate to four decimal places and then graph the nth roots of unity for (a) $n = 5$, (b) $n = 7$, (c) $n = 8$, (d) $n = 9$, and (e) $n = 10$.

3. For each of the following pairs z, n, approximate to four decimal places, and then graph the nth roots of z.

    a) $z = 2, n = 3$ b) $z = 2 + i, n = 5$ c) $z = -2 + i, n = 4$
    d) $z = -7, n = 5$ e) $z = 9i - 3, n = 6$ f) $z = -1000000, n = 6$

4. Find exactly all of the solutions to the following equations.

    a) $z^2 = -1$ b) $z^2 = i$ c) $z^4 = 16$
    d) $z^3 = -1$ e) $z^3 = -i$ f) $z^4 = -16$
    g) $z^6 = -1$ h*) $z^5 = 32$ i) $z^8 = 1/256$.

5. Approximate all of the solutions to the following equations.

    a) $z^2 = -7$ b) $z^2 = 2 + i$ c) $z^4 = 4 + 3i$
    d) $z^3 = 2 - i$ e) $z^5 = 3 - i$ f) $z^4 = 4 - 3i$
    g) $z^{11} = -1$ h) $z^5 = 32$ i) $z^8 = 8$.

6. (*) Prove Theorem 2.6.1. Hint: First show that all the nth roots are on the circle centered at the origin with radius $|z|^{1/n}$. (Explain why it is enough to show that all the nth roots of z have absolute value $|z|^{1/n}$.) Then show that the nth roots are evenly spaced about this circle.

7. (*) We have seen that a non-zero complex number z has n n^{th} roots. A nth root w of z for which all the nth roots of z may be written in the form w^m (m an integer) is called a **primitive n^{th} root**. Let z be a non-zero complex number.

    a) Show that if $z = Re^{i\theta}$, then $w = R^{1/n}e^{i\theta/n}$ is a primitive of z.

2. COMPLEX NUMBERS

b) Show that z has one primitive square root.

c) Show that z has two primitive cube roots.

d) Show that z has two primitive fourth roots.

e) Show that z has four primitive fifth roots.

f) Show that z has two primitive sixth roots.

g) Show that z has six primitive seventh roots.

h) (N) Show that z has $\phi(n)$ primitive n^{th} roots where $\phi(n)$ is the Euler *phi-function* ($\phi(n)$ is the number of integers between 1 and n (inclusive) which are relatively prime to n).

Note that a given primitive root, like any complex number, has infinitely many possible arguments. For this problem pick arguments θ with $0 \leq \theta < 2\pi$.

Polynomials

*How can it be that mathematics,
being after all a product of human thought
independent of experience,
is so admirably adapted to the objects of reality?*

Albert Einstein

*It has often happened that notations have been
inseparable from major theoretical advances.*

A. Weil

I do not know.

J. L. Lagrange

Chapter 3

Polynomials

In this chapter we recall the basic definitions involved in the study of polynomials, and begin our study concentrating on the operation of division. It is assumed that the reader can add, subtract, multiply, and (long) divide polynomials.

Sometimes students balk when presented a list of a dozen or more definitions (as we do in the next section), but as teachers we need to remind them how the words they know control the thoughts they think. We think in words. We write in words. We whine and complain in words. Changing those words not only allows us to discuss new things, but changes how we view them. In this chapter we get ready to think deeply about polynomials, so we begin with polynomial words.

3.1 Basic Definitions

An expression that can be written in the **standard form**

$$P(x) = a_0 x^n + a_1 x^{n-1} + a_2 x^{n-2} + \ldots + a_{n-1} x + a_n,$$

where n is a positive integer and $a_0, a_1, \ldots, a_n$ are any constants, is called a **polynomial in x** or just a **polynomial**. The constants $a_0, a_1, \ldots, a_n$ are called the **coefficients**, x is called the **variable**, and the single **monomials**

$$a_0 x^n, a_1 x^{n-1}, a_2 x^{n-2}, \ldots, a_n$$

are called the **terms** of the polynomial.

When no confusion should arise, we will write P instead of $P(x)$. If a_0 is not zero, then the polynomial is of **degree** n, a_0 is the **leading coefficient**, and $a_0 x^n$ is the **leading term**. Finally, a_n is called the **constant term** (note $a_n = P(0)$). It is convenient to consider a (non-zero) constant c as a polynomial of degree zero ($c = cx^0$).

3. POLYNOMIALS

Two polynomials are **identical** (or identically equal) if they always take on the same values for every value of x. This happens if and only if they are equal term by term; that is,

$$a_0 x^n + a_1 x^{n-1} + \ldots + a_n = b_0 x^n + b_1 x^{n-1} + \ldots + b_n$$

if and only if

$$a_0 = b_0, a_1 = b_1, \ldots, a_n = b_n.$$

If every coefficient of a polynomial is zero, then the polynomial is **identically vanishing** and can be replaced by 0. No degree is assigned to an identically vanishing polynomial (but see problem 5).

Polynomials of degree one (two, three, four or five) are called **linear** (**quadratic**, **cubic**, **quartic** or **quintic** respectively.) In many older texts quartic polynomials are called **biquadratic** polynomials. If the leading coefficient of a polynomial is one, then it is called a **monic polynomial**.

If the coefficients of a polynomial are all integers, then it is called an **integral polynomial**. Finally, if it has all rational coefficients, then it is called a **rational polynomial**.

Exercises 3.1:

1. Prove there is only one identically vanishing polynomial. It is called the **Zero Polynomial**. Every other polynomial is said to be a **non-zero polynomial**.

2. Let the nonzero polynomials P, Q have degrees n, m respectively. Show that the degree of PQ is $n+m$ (PQ is the product of P and Q).

3. Let the nonzero polynomials P, Q have degrees n, m respectively. Show that the degrees $P+Q$ and $P-Q$ are either undefined or at most $\max(n, m)$, and that these degrees are both $\max(n, m)$ if $m \neq n$. Hint: The degree is decided by the leading terms only identically vanishing polynomials have undefined degree.

4. Let P be a nonzero polynomial of degree n. Show P^m has degree mn.

5. (*) Show that if we define the degree of the zero polynomial to be negative infinity, then the word nonzero is not necessary in the previous three problems.

6. Show that the degree of the zero polynomial cannot be an integer (positive, negative, or zero).

7. Why is it that $3x^{-1}+2$ and $(3x+2)/(4x+6)$ are not polynomials?

8. Explain why the following is not a polynomial.

$$e^x = 1 + x + \frac{x^2}{2} + \frac{x^3}{3!} + \frac{x^4}{4!} + \frac{x^5}{5!} + \cdots$$

9. (A) Let R be a ring. Define $R[x]$ to be all polynomials with coefficients in R (that is, $a_0, a_1, \ldots, a_n$ are elements of R).

   a) Show $R[x]$ is a ring. $R[x]$ is called the **ring of polynomials of R**. (Here we assume, as usual, that x is a variable.)
   b) Prove $R[x]$ is a field if and only if R is the trivial ring $\{1\}$.
   c) If R is a domain, show that the results of problems 1-4 also hold in $R[x]$.
   d) Give examples of rings R for which the results of problems 2 and 4 do not hold in the polynomial rings $R[x]$.

3.2 The Remainder and Factor Theorem

In secondary school we learned how to divide, using long division, a polynomial $P(x)$ by another polynomial $Q(x)$ to get a quotient and a remainder. For example, we have the following.

$$
\begin{array}{r}
x+1 \\
x^2+2x+1 \overline{\smash{)}\ x^3+3x^2+6x+3} \\
-x^3-2x^2-x \\
\hline
x^2+5x+3 \\
-x^2-2x-1 \\
\hline
3x+2
\end{array}
$$

By carefully describing how this division is performed in general, we can prove the following result.

Theorem 3.2.1 (Division Algorithm). *Let $P(x)$ and $D(x)$ be polynomials with $D(x)$ not the zero polynomial. There exist <u>unique</u> polynomials $Q(x)$ and $R(x)$ for which*

$$P(x) = Q(x)D(x) + R(x)$$

*where either $R(x) = 0$, or the degree of $R(x)$ is <u>less than</u> the degree of $D(x)$. The polynomial $Q(x)$ is called the **quotient** and $R(x)$ is the **remainder**.*

In a later section we will discuss a method for performing this polynomial division. First let us explore two consequences of the Division Algorithm: the Remainder Theorem and the Factor Theorem.

3. POLYNOMIALS

> **Theorem 3.2.2 (Remainder Theorem).** *If a polynomial $P(x)$ is divided by $x - c$, then the remainder is $P(c)$.*

Example 3.2.1. *Without dividing find the remainder when $x^3 - 6x^2 - 8x + 7$ is divided by $x - 2$.*

Solution: By the Remainder Theorem the remainder is $P(2) = -25$. ∎

A polynomial $D(x)$ is a **factor** (or **divisor**) of $P(x)$ if and only if $D(x)$ exactly divides $P(x)$, that is, if and only if there is another polynomial $Q(x)$ such that $P(x) = D(x)Q(x)$. An important special case of the Remainder Theorem is the following.

> **Theorem 3.2.3 (Factor Theorem).** *A polynomial $P(x)$ has the factor $x - c$ if and only if $P(c) = 0$.*

Example 3.2.2. *Without dividing, show that $x - 7$ divides $x^3 - 6x^2 - 8x + 7$.*

Solution: We need only to show that $P(7) = 0$ by the Factor Theorem. This is easily checked. ∎

Exercises 3.2:

1. Assuming the Division Algorithm, prove the Remainder Theorem.

2. Show that the Factor Theorem is a direct consequence of the Remainder Theorem.

3. Show that in the Division Algorithm, the degree of P is the degree of Q plus the degree of D.

4. (*) Prove the Division Algorithm. Hint: Show that if the degree of $D(x)$ is less than or equal to the degree of $R(x)$, then by changing $Q(x)$, the degree of $R(x)$ may be decreased. To show the uniqueness of $Q(x)$ and $R(x)$, let Q_1, R_1 and Q_2, R_2 satisfy the conditions of the theorem. Then

$$P(x) - P(x) = (Q_1(x)D(x) + R_1(x)) - (Q_2(x)D(x) + R_2(x)),$$

   so $(Q_1 - Q_2)D = R_1 - R_2$. Think about the degrees of the polynomials in this last equation.

5. Let the nonzero polynomials P, Q have degrees n, m respectively. Show that if Q divides P, then the degree of P/Q is $n-m$.

3.2. The Remainder and Factor Theorem

6. Show that if $D(x)$ is a factor of $P(x)$ and $a \neq 0$, then $aD(x)$ is also a factor of $P(x)$.

7. Use the Factor Theorem to show the following without dividing.

    a) $x + 2$ divides $x^4 - 3x^3 - 6x^2 + 3x - 10$

    b) $x - 1$ divides $8x^9 - 4x^7 + 6x^3 - 11x + 1$

    c) $x - 7$ divides $x^3 - 6x^2 - 8x + 7$

    d) $2x - 1$ divides $16x^4 - 12x^2 + 6x - 1$

    e) $3x + 2$ divides $27x^3 + 18x^2 + 6x + 4$

    f) $x - 2$ divides $x^{103} - x^{102} - 4x^{100}$

    g) $x^2 - 3x + 2$ divides $x^4 - x^3 - 7x^2 + 13x - 6$

    Hint: For (d) and (e) use the previous problem. For (g) factor $x^2 - 3x + 2$.

8. Suppose we have divided $P(x)$ by $D(x) \neq 0$ to find the quotient $Q(x)$ and remainder $R(x)$. Show that if a is any non-zero number, then the quotient and remainder upon division of $P(x)$ by $aD(x)$ are $Q(x)/a$ and $R(x)$ respectively.

9. Without dividing find the remainder when

    a) $x^4 - 3x^3 - 6x^2 + 3x - 10$ is divided by $x - 2$

    b) $8x^9 - 4x^7 + 6x^3 - 11x + 1$ is divided by $x + 1$

    c) $x^3 - 6x^2 - 8x + 7$ is divided by $x - 2$

    d) $16x^4 - 12x^3 + 6x - 1$ is divided by $2x + 1$

    e) $27x^3 + 18x^2 + 6x + 4$ is divided by $3x + 2$

    f) $x^{103} - x^{102} - 4x^{100}$ is divided by $x - 1$

    g) $x^4 - x^3 - 7x^2 + 13x - 6$ is divided by $x + 3$

    h) $x^{103} - x^{102} - 4x^{100}$ is divided by $x + i$

    i) $x^4 - 3x^3 - 6x^2 + 3x - 10$ is divided by $x - 1 - i$

    Hint: For (d) and (e) use the previous problem.

10. Find both the quotient and remainder when

    a) $x^4 + 3x^3 - 10x^2 - 7x + 3$ is divided by $x^2 - 3x + 2$

    b) $x^3 - 6x^2 + 4x - 8$ is divided by $x^2 - 2$

    c) $2x^4 + x^3 + 2x + 1$ is divided by $x^2 - x + 1$

    d) $3x^5 + 3x^3 - 5x^2 + 4x - 8$ is divided by $x^2 + x + 1$

11. Prove the following.

3. POLYNOMIALS

> **Theorem 3.2.4.** *The binomial $x - 1$ divides $P(x)$ if and only if the sum of the coefficients of $P(x)$ is zero.*

12. Prove the following.

> **Theorem 3.2.5.** *The binomial $x + 1$ divides $P(x)$ if and only if the sum of the coefficients of the odd power terms of $P(x)$ is equal to the sum of the coefficients of the even power terms.*

13. Let n be a positive integer. What are the conditions on n and a under which $x^n + a^n$ is divisible by $x + a$? by $x - a$?

14. Prove the following extension of the Factor Theorem.

> **Corollary 3.2.6.** *The remainder upon dividing $P(x)$ by $ax - b$ with $a \neq 0$, is $P(\frac{b}{a})$.*

15. (**) Show that $(x+1)^n - x^n - 1$ is divisible by $x^2 + x + 1$ only if n is an odd number which is not divisible by 3. Hint: What is $(x-1)(x^2 + x + 1)$?

16. (A) Prove that the Factor Theorem is valid in every ring R.

17. (A) Let R be a field. The Division Algorithm holds for R. Use this fact to prove that the Remainder Theorem and the Factor Theorem also hold for $R[x]$.

3.3 Synthetic Division

In this section we present a marvelously simple to use, but awkward to describe algorithm for dividing a polynomial P by $x - c$.

> **Theorem 3.3.1 (Synthetic Division Algorithm).** Let $P(x) = (x - c)Q(x) + r$ where
> $$P(x) = a_0 x^n + a_1 x^{n-1} + \ldots + a_n$$
> and
> $$Q(x) = b_0 x^{n-1} + b_1 x^{n-2} + \ldots + b_{n-1}.$$
> Then $b_0, b_1, \ldots, b_{n-1}$ and r can be found recursively as follows:
> $$b_0 = a_0, b_1 = a_1 + cb_0, b_2 = a_2 + cb_1, \ldots, b_{n-1} = a_{n-1} + cb_{n-2}, r = a_n + cb_{n-1}.$$

3.3. Synthetic Division

When using this algorithm it is very helpful to arrange our work as follows

$$
\begin{array}{c|ccccc}
c & a_0 & a_1 & a_2 & \ldots & a_{n-1} & a_n \\
& & b_0 c & b_1 c & \ldots & b_{n-2} c & b_{n-1} c \\
\hline
& b_0 & b_1 & b_2 & \ldots & b_{n-1} & r
\end{array}
$$

In the first line all the coefficients of $P(x)$ are written (*without* omitting the coefficients that are zero). The third line is begun with $b_0 = a_0$ (a_0 is "brought down"). Then b_0 is multiplied by c, and their product is written on the second line. Next this product is added to a_1, and the sum is placed on the third line. We next repeat this process with b_1 instead of b_0: first b_1 is multiplied by c; their product is placed on the second line and added to a_2; then the sum is placed on the third line. We continue this process until we find r in the last column on the third line. Notice how c and r are separated from the other entries for ease of reading.

That was a bit hard to follow right? Let's do this slowly with an example.

Example 3.3.1. *Find the quotient and remainder upon dividing*

$$3x^6 - 7x^5 + 12x^4 + 6x^2 - 4x + 8 \text{ by } x - 2.$$

Solution: Start by writing

$$
\begin{array}{c|ccccccc}
2 & 3 & -7 & 12 & 0 & 6 & -4 & 8 \\
& & & & & & & \\
\hline
& 3 & & & & & &
\end{array}
$$

Notice we have written the coefficients of the polynomial on the first line *without* omitting the coefficient 0 of x^3, and "brought down" the leading coefficient 3. Now 3 is multiplied by 2 and added to -7, with the results being placed as follows:

$$
\begin{array}{c|ccccccc}
2 & 3 & -7 & 12 & 0 & 6 & -4 & 8 \\
& & 6 & & & & & \\
\hline
& 3 & -1 & & & & &
\end{array}
$$

Next -1, the rightmost entry on the bottom line, is multiplied by 2 and added to 12, with the results again placed as follows:

$$
\begin{array}{c|ccccccc}
2 & 3 & -7 & 12 & 0 & 6 & -4 & 8 \\
& & 6 & -2 & & & & \\
\hline
& 3 & -1 & 10 & & & &
\end{array}
$$

3. POLYNOMIALS

We continue this process of multiplying the rightmost entry on the last line by $c = 2$; placing the result on the second line one column to the right; and adding until we find the remainder r. The completed table is below.

$$\begin{array}{r|rrrrrr|r} 2 & 3 & -7 & 12 & 0 & 6 & -4 & 8 \\ & & 6 & -2 & 20 & 40 & 92 & 176 \\ \hline & 3 & -1 & 10 & 20 & 46 & 88 & 184 \end{array}$$

So we have found that the quotient is

$$3x^5 - x^4 + 10x^3 + 20x^2 + 46x + 88$$

and the remainder is $r = 184$. When using synthetic division, we are dividing by a linear term, so the degree of quotient is *always* one less than the degree of the polynomial we started with. ∎

Example 3.3.2. *Divide $3x^8 - 20x^6 - 6x^5 - 62x^4 + 56x^3 - 4x + 17$ by $x + 3$.*

Solution: The work is shown as follows.

$$\begin{array}{r|rrrrrrrr|r} -3 & 3 & 0 & -20 & -6 & -62 & 56 & 0 & -4 & 17 \\ & & -9 & 27 & -21 & 81 & -57 & 3 & -9 & 39 \\ \hline & 3 & -9 & 7 & -27 & 19 & -1 & 3 & -13 & 56 \end{array}$$

So the quotient is

$$3x^7 - 9x^6 + 7x^5 - 27x^4 + 19x^3 - x^2 + 3x - 13$$

and the remainder is $r = 56$. ∎

Exercises 3.3:

1. Use synthetic division to find the quotient and remainder if the first polynomial is divided by the second.

   a) $3x^4 - 6x^3 - 4x^2 + 5x + 7$, $x - 2$
   b) $-x^4 + 7x^3 + 6x^2 + 7$, $x + 3$
   c) $-2x^6 + 10x - 37$, $x - 1$
   d) $3x^4 + 8x^3 - 2x^2 - 10x + 4$, $x + 2$
   e) $x^4 + 3x^3 - 5x^2 + 2x + 1$, $x + 1 + i$
   f) $3x^4 - 6x^3 - 4x^2 + 5x + 7$, $-x + 1$

g) $-x^4 + 7x^3 + 6x^2 + 7$, $\quad -x + 3$

h) $x^4 + 3x^3 - 5x^2 + 2x + 1$, $\quad x + (i - 1)$

i) $3x^4 + 8x^3 - 2x^2 - 10x + 4$, $\quad 2x + 1$

j) $-2x^6 + 10x - 37$, $\quad 2x - 1$

2. Let n be a positive integer. What is the quotient when $x^n - r^n$ is divided by $x - r$?

3. (*) Let $r, r_1, r_2, \ldots, r_{n-1}$ be the zeros of $x^n - r^n = 0$. Show that

$$(r - r_1)(r - r_2) \ldots (r - r_{n-1}) = nr^{n-1}.$$

Hint: Use the result of the previous exercise.

4. The nth roots of 1 form a regular n-gon (Theorem 2.6.1, page 29) inscribed in the unit circle with a vertex at 1. What is the product of the lengths of the $n-1$ diagonals from 1 to the other vertices? (See Figure 3.1 for an example with $n = 6$).

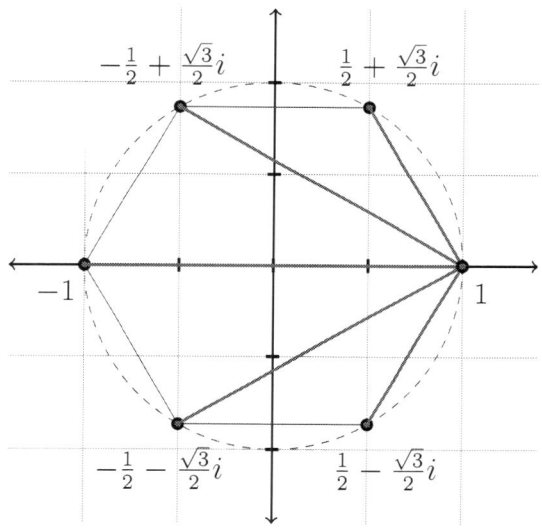

Figure 3.1: The Diagonals to 1 (Problem 4, Section 3.3)

5. In programming a computer an efficient way to calculate

$$P(c) = a_0 c^n + a_1 c^{n-1} + \ldots + a_n$$

is to rearrange the work as follows

$$(\ldots(((a_0 c + a_1)c + a_2)c + a_3)\ldots)c + a_n.$$

Show that the work in calculating $P(c)$ in this manner is exactly the same as the work involved in *synthetically* dividing $P(x)$ by $x - c$ (and is often called **Horner's Rule**). Compare this result with the Remainder Theorem.

3. POLYNOMIALS

6. Write a subroutine which divides $P(x)$ by $x - c$. The routine should have as input: the integer n (which is the degree of $P(x)$); the array $A(0), \ldots, A(n)$ (which are the coefficients of P); and the real number c. The output should be: an array $B(0), \ldots, B(n-1)$ (which are the coefficients of the quotient) and the number R (which is $P(c)$, i.e., the remainder).

7. (*) Prove the Synthetic Division Algorithm. Hint: Use long division to perform the division of example 3.3.1. Where do the numbers which occur in the synthetic division procedure occur during long division?

3.4 The Greatest Common Divisor

Let P and Q be polynomials. Recall that P divides Q (or P is a divisor of Q, or P is a factor of Q) if and only if there is a third polynomial R such that $Q = P \cdot R$. If P divides Q, then aP divides Q for all non-zero constants a because $Q = PR$ implies $Q = (aP)(R/a)$.

A **common divisor** of P and Q is a polynomial which divides both P and Q. Common divisors are also called **common factors**. Clearly a common divisor of P and Q must have a degree less than or equal to the degrees of P and Q, so in the set of common divisors of P and Q there is a maximum degree. Any common divisor of P and Q with the maximum degree is call a **greatest common divisor** and is denoted **gcd**(P, Q).

Example 3.4.1. *Find the greatest common divisor of the following two polynomials.*

$$x^4 - 1 = (x+1)(x-1)(x+i)(x-i)$$

$$x^3 - 2x^2 + x - 2 = (x-2)(x+i)(x-i)$$

Solution: These polynomials have both $(x+i)$ and $(x-i)$ in common. So the greatest common divisor is the product of these: $x^2 + 1 = (x+i)(x-i)$ or any rational number multiple of this. For polynomials, the "greatest" in "greatest common divisor" means of largest degree. ■

Example 3.4.2. *Find the greatest common divisor of $24x^4 - 52x^3 + 42x^2 - 15x + 2$ and $36x^4 - 84x^3 + 73x^2 - 28x + 4$.*

Solution: These polynomials factor nicely as

$$24\left(x - \frac{1}{2}\right)^3 \left(x - \frac{2}{3}\right) \quad \text{and} \quad 36\left(x - \frac{1}{2}\right)^2 \left(x - \frac{2}{3}\right)^2$$

respectively, so the greatest common divisor is

$$\left(x-\frac{1}{2}\right)^2\left(x-\frac{2}{3}\right) = x^3 - \frac{5}{3}x^2 + \frac{11}{12}x - \frac{1}{6}$$

or any rational number multiple of this. Since this answer has fractional coefficients, we might multiply this answer by 12 to get the following result with integer coefficients.

$$12\left(x-\frac{1}{2}\right)^2\left(x-\frac{2}{3}\right) = 12x^3 - 20x^2 + 11x - 2.$$

∎

These two examples show that finding the greatest common divisor is easy if you have factored both polynomials. We address factoring polynomials in the next chapter, but this is often very difficult to do (at least by hand). However, there is a very old method that will always work, and which does not require factorization—let's take a look.

The Euclidean Algorithm

Let P and Q be polynomials. One approach to finding $\gcd(P, Q)$ is based on the Division Algorithm. Begin by dividing P by Q, finding a quotient q_1 and remainder r_1:

$$P(x) = q_1(x)Q(x) + r_1(x).$$

Any common divisor of P and Q must also be a common divisor of Q and r_1 (exercise 1), so

$$\gcd(P, Q) = \gcd(Q, r_1).$$

This is a step forward because the division reduces the degrees of the polynomials we are working with. Notice that we do not really care what the quotient q_1 is once we have the remainder.

Next we divide Q by r_1, finding a remainder r_2 with degree less than r_1; and again we have

$$\gcd(P, Q) = \gcd(Q, r_1) = \gcd(r_1, r_2).$$

Next we divide r_1 by r_2, finding a remainder r_3 with even lower degree, and so on. This process must eventually result in a constant remainder because the polynomials $r_1, r_2, r_3, \ldots$ are of decreasing degree. Let this constant remainder be denoted by r_{n+1}.

3. POLYNOMIALS

If the successive quotients are denoted by $q_1, q_2, \ldots, q_{n+1}$ we have

$$\begin{aligned}
P(x) &= q_1(x)Q(x) + r_1(x) \\
Q(x) &= q_2(x)r_1(x) + r_2(x) \\
r_1(x) &= q_3(x)r_2(x) + r_3(x) \\
r_2(x) &= q_4(x)r_3(x) + r_4(x) \\
&\ldots \\
r_{n-2}(x) &= q_n(x)r_{n-1}(x) + r_n(x) \\
r_{n-1}(x) &= q_{n+1}(x)r_n(x) + r_{n+1}
\end{aligned}$$

Using these equalities we can see that every common factor of P and Q is a factor of all the r_i's, and that every common factor of two consecutive r_i's is a common factor of P and Q. Thus any common factor of P and Q must divide the constant remainder r_{n+1}, so r_{n+1} either zero and $\gcd(P,Q) = r_n$, or r_{n+1} is not zero and $\gcd(P,Q) = q$. **Either way, the last nonzero remainder is the greatest common divisor of P and Q.**

It is important to note that at any time during the Euclidean algorithm we may multiply any of the polynomials involved by a non-zero constant. For example, we might multiply by a constant to avoid working with fractions, or divide by a constant to reduce the size of the coefficients.

That description was quite a mouthful—I could very well understand a poor reader getting lost; I might have. Let's try to clear this up with an example.

Example 3.4.3. *Find the greatest common divisor of P and Q where*

$$\begin{aligned}
P(x) &= x^6 - 4x^5 + 3x^4 - 4x^3 + 6x^2 - 16x + 8, \text{ and} \\
Q(x) &= x^5 - 5x^4 + 5x^3 + 3x^2 - 6x + 2.
\end{aligned}$$

Solution: The first step in the Euclidean algorithm is to divide P by Q. The division is performed using long division as follows.

$$\begin{array}{r}
x + 1 \\
x^5 - 5x^4 + 5x^3 + 3x^2 - 6x + 2 \overline{\smash{)}\, x^6 - 4x^5 + 3x^4 - 4x^3 + 6x^2 - 16x + 8} \\
\underline{- x^6 + 5x^5 - 5x^4 - 3x^3 + 6x^2 - 2x } \\
x^5 - 2x^4 - 7x^3 + 12x^2 - 18x + 8 \\
\underline{- x^5 + 5x^4 - 5x^3 - 3x^2 + 6x - 2} \\
3x^4 - 12x^3 + 9x^2 - 12x + 6
\end{array}$$

The first remainder is

$$r_1 = 3x^4 - 12x^3 + 9x^2 - 12x + 6.$$

3.4. The Greatest Common Divisor

We now have to divide Q by r_1. Before we proceed, we divide r_1 by 3 to simplify our computations. The next step is then as follows.

$$
\begin{array}{r}
x - 1 \\
x^4 - 4x^3 + 3x^2 - 4x + 2 \overline{\smash{\big)}\, x^5 - 5x^4 + 5x^3 + 3x^2 - 6x + 2} \\
\underline{-x^5 + 4x^4 - 3x^3 + 4x^2 - 2x} \\
-x^4 + 2x^3 + 7x^2 - 8x + 2 \\
\underline{x^4 - 4x^3 + 3x^2 - 4x + 2} \\
-2x^3 + 10x^2 - 12x + 4
\end{array}
$$

Every coefficient in the remainder is divisible by -2, so we may take for the second remainder:

$$r_2 = x^3 - 5x^2 + 6x - 2.$$

We continue by dividing r_1 by r_2.

$$
\begin{array}{r}
x + 1 \\
x^3 - 5x^2 + 6x - 2 \overline{\smash{\big)}\, x^4 - 4x^3 + 3x^2 - 4x + 2} \\
\underline{-x^4 + 5x^3 - 6x^2 + 2x} \\
x^3 - 3x^2 - 2x + 2 \\
\underline{-x^3 + 5x^2 - 6x + 2} \\
2x^2 - 8x + 4
\end{array}
$$

So we may use $x^2 - 4x + 2$ as r_3. Finally dividing r_2 by r_3 we get the following.

$$
\begin{array}{r}
x - 1 \\
x^2 - 4x + 2 \overline{\smash{\big)}\, x^3 - 5x^2 + 6x - 2} \\
\underline{-x^3 + 4x^2 - 2x} \\
-x^2 + 4x - 2 \\
\underline{x^2 - 4x + 2} \\
0
\end{array}
$$

The remainder is zero, so the previous remainder,

$$x^2 - 4x + 2,$$

is the greatest common divisor of P and Q. In fact we find

$$
\begin{aligned}
P(x) &= (x^2 - 4x + 2)(x^4 + x^2 + 4) \\
Q(x) &= (x^2 - 4x + 2)(x^3 - x^2 - x + 1).
\end{aligned}
$$

■

3. POLYNOMIALS

Of course Euclid did not use polynomials a couple thousand years ago; he did this process with integers. Perhaps you have done this with integers too? In many older texts the greatest common divisor is called the **highest common factor** or the **greatest common measure**.

In Appendix B.1 we define the resultant of two polynomials, which provides a way of finding whether two polynomials have a non-trivial greatest common divisor without dividing (and to even find the degree of that common divisor).

Before the exercises, let us have one more particularly easy example.

Example 3.4.4. *Find the greatest common divisor of P and Q where*

$$P(x) = x^6 + x^5 + x^4 + x^3 + x^2 + x + 3, \text{ and}$$
$$Q(x) = x^5 + x^4 + x^3 + x^2 + x + 1.$$

Solution: We can see by inspection that the quotient when $P(x)$ is divided by $Q(x)$ is just x and the remainder is just 3. The next remainder (the remainder when we divide a polynomial by 3) will be zero, so $\gcd(P,Q) = 3$ (or any non-zero rational multiple thereof). It looks nicer to choose $\gcd(P,Q) = 1$. ∎

Exercises 3.4:

1. Suppose we divide $P(x)$ by $Q(x)$ to get a remainder $r(x)$ (with P and Q non-zero.)

    a) Prove that any common divisor of $P(x)$ and $Q(x)$ divides $r(x)$

    b) Prove that any common divisor of $Q(x)$ and $r(x)$ divides $P(x)$.

2. Find the greatest common divisor of the following pairs of polynomials.

    a) $P(x) = x^4 + x^3 + 4x^2 + 3x + 3$, $\quad Q(x) = x^3 + 2x^2 + 3x + 6$

    b) $P(x) = x^4 + 8x^3 + 10x^2 + 9x + 2$, $\quad Q(x) = x^3 + 2x^2 + 2x + 1$

    c) $P(x) = x^4 + 5x^3 + 9x^2 + 8x + 4$, $\quad Q(x) = x^3 + 6x^2 + 9x + 2$

    d) $P(x) = x^4 + x^3 + x^2 + x + 1$, $\quad Q(x) = x^6 + x^5 + x^4 + x^3 + x^2 + x + 1$

    e) $P(x) = x^4 + 3x^3 + 3x^2 + 2x$, $\quad Q(x) = x^6 + x^5 + x^4 + x^3 + x^2 + x + 1$

    f) $P(x) = x^4 + 2x^3 + 3x^2 + 4x + 5$, $\quad Q(x) = x^3 + 2x^2 + 3x$

3. (*) If p, q are integers it is possible to find integers a, b such that $ap + bq = \gcd(a, b)$. Show that the same is true for polynomials—that is, given $P(x)$ and $Q(x)$, show that we may find polynomials $A(x)$ and $B(x)$ so that $A(x)P(x) + B(x)Q(x) = \gcd(P(x), Q(x))$.

4. (*) Show that the greatest common divisor is unique (except for multiplication by a constant). That is, prove that if $D_1(x)$ and $D_2(x)$ are greatest common divisors of $P(x)$ and $Q(x)$, then $D_1(x)/D_2(x)$ is a non-zero constant.

3.4. The Greatest Common Divisor

5. ▨ Write a program to find the greatest common divisor of two polynomials.

6. (*) Let a, b be relatively prime integers and $P(x)$ be a polynomial with integral coefficients. Show that if $P(a/b) = 0$, then $a - b$ divides $P(1)$. Hint: $bP(x)$ has the factor $(bx - a)$.

7. Show that we can complete a version of Euclid's Algorithm on P and Q without ever doing polynomial division. Hint: If the polynomials have the same degree, then subtract, and continue the process with Q and $P - Q$. Otherwise, multiply the polynomial with lower degree by whatever necessary power of x, and then subtract. Either way, the difference plays the same role as the remainder in Euclid's Algorithm.

Zeros of Polynomials

Nature's great book is written in mathematical symbols.

Galileo

Mathematics for mathematics' sake. People have been shocked by this formula and yet it is as good as life for life's sake, if life is but misery.

Poincaré

If a man who cannot count finds a four-leaf clover, is he entitled to happiness?

S. J. Lec

Chapter 4

Zeros of Polynomials

In this chapter we will just begin the study of the zeros of polynomials. After a few definition we will discuss the Fundamental Theorem of Algebra; the complex zeros of polynomials with real coefficients; techniques for finding multiple zeros; and Lagrange's Interpolation Formula. In the next chapters we will build on this base.

4.1 Polynomial Equations

LET $P(x)$ BE A NON-ZERO POLYNOMIAL. A **zero** or **root** of $P(x)$ is a solution of the **polynomial equation $P(x) = 0$**. The process of finding these zeros is called **solving** the equation $P(x) = 0$ or factoring $P(x)$.

One of the main approaches to factoring a polynomial is to divide an concur (pun intended). We try to find the zeros one at a time, and then divide them out, leaving us a hopefully simpler polynomial to factor. Recall that the Factor Theorem states that if r is a zero of P, then $(x-r)$ divides $P(x)$. It is useful to extend this as follows.

Theorem 4.1.1 (Extended Factor Theorem). *If $r_1, r_2, \ldots, r_m$ are distinct zeros of P, then $(x - r_1)(x - r_2)\ldots(x - r_m)$ divides $P(x)$.*

Now suppose we want to solve $P(x) = 0$ and we are given (or have found) some of its zeros $r_1, r_2, \ldots, r_m$. Using this theorem we know that there is another polynomial $Q(x)$ such that
$$P(x) = (x - r_1)(x - r_2)\ldots(x - r_m)Q(x)$$
Any additional zero of $P(x)$ must also be a zero of $Q(x)$. This means we can find the rest of the zeros of $P(x)$ by finding the zeros of $Q(x)$.

4. Zeros of Polynomials

Example 4.1.1. *Solve the following quartic equation given that two of the zeros are 1 and 3.*
$$P(x) = x^4 - 7x^3 + 11x^2 + 7x - 12 = 0.$$

Solution: Because 1 and 3 are zeros, we know that $P(x) = (x-1)(x-3)Q(x)$, and that the other zeros of $P(x)$ are the zeros of $Q(x)$. We may find $Q(x)$ by dividing $P(x)$ first by $x-1$ and then by $x-3$ as follows.

$$\begin{array}{r|rrrr|r} 1 & 1 & -7 & 11 & 7 & -12 \\ & & 1 & -6 & 5 & 12 \\ \hline & 1 & -6 & 5 & 12 & 0 \end{array}$$

$$\begin{array}{r|rrr|r} 3 & 1 & -6 & 5 & 12 \\ & & 3 & -9 & -12 \\ \hline & 1 & -3 & -4 & 0 \end{array}$$

So $P(x) = (x-1)(x+3)(x^2 - 3x - 4)$. The quadratic $x^2 - 3x - 4$ has the zeros -1 and 4 (these may be found either by inspection or by the quadratic formula); hence $P(x)$ has the zeros 1, -1, 3, and 4. ∎

Example 4.1.2. *Find a polynomial $P(x)$ with the zeros 2, 3, and -5.*

Solution: By the Extended Factor Theorem every polynomial with the zeros 2, 3 and -5 is divisible by:
$$(x-2)(x-3)(x+5) = x^3 - 19x + 30.$$

We offer this cubic as the desired polynomial $P(x)$. Notice that any multiple of $P(x)$ will also work. ∎

If a polynomial P has degree n, then every divisor of P has degree less than or equal to n. Using this and the Extended Factor Theorem we can easily prove the following.

Corollary 4.1.2. *A non-zero polynomial of degree n has at most n distinct zeros.*

Corollary 4.1.3. *Let $P(x) = a_0 x^n + a_1 x^{n-1} + \ldots + a_n$. If $P(x) = 0$ for more than n distinct values of x, then $P(x)$ is identically zero.*

4.1. Polynomial Equations

Exercises 4.1:

1. In the definition of a polynomial $P(x)$, x is called the "variable." In the definition of the polynomial equation $P(x) = 0$, x is called the "unknown." What is the difference between a variable and an unknown? Hint: Can the unknown in a polynomial equation take on any value?

2. We have discussed the "polynomial $P(x) = 0$," the "identity $P(x) = 0$," and the "equation $P(x) = 0$." What is the difference (if any) between these three concepts?

3. Use the Extended Factor Theorem to find polynomials with the given zeros.

   a) 2, −3 and 4.
   b) 1, −1, 2, −2, 3 and 3.
   c) 2, $1 + 3i$ and $1 - 3i$.
   d) i, $-i$, $1 + i$ and $1 - i$.
   e) −1, i, and $-i$.

4. Find all the zeros of the following polynomials (each has the given zeros).

   a) $x^3 + 3x^2 - 4x - 12$; zero 2.
   b) $x^3 - x^2 - 11x - 4$; zero 4.
   c) $4x^4 - 11x^2 + 9x - 2$; zeros 1, 1/2.
   d) $x^5 - 4x^3 - x^2 + 4$; zeros 1, −2, 2.

5. Prove the following.

   Theorem 4.1.4 (Identity Theorem). *Let $P(x)$ and $Q(x)$ be polynomials with degree less than m. If $P(x) = Q(x)$ for m distinct values of x, then P and Q are identical polynomials.*

   Hint: Apply Corollary 4.1.3 to the polynomial $P(x) - Q(x)$.

6. Prove the Extended Factor Theorem. Hint: By the Factor Theorem we have that $P(x) = (x - r_1)Q_1(x)$. Now using the fact that the r_i are distinct, show that $Q_1(r_2) = 0$ so $(x - r_2)$ divides Q_1...

7. Prove Corollary 4.1.2. Hint: Degrees.

8. Prove Corollary 4.1.3. Hint: What is the degree of the polynomial?

9. Let c be any number and n any positive integer. Show that c has at most n n^{th} roots. Hint: What is the definition of an nth root? Use Corollary 4.1.2.

10. (N) Show that $x^3 - x$ has 6 zeros in the ring $Z/6Z$ (the integers modulo 6). This shows the Extended Factor Theorem is not valid for $Z/6Z$.

4. Zeros of Polynomials

4.2 The Fundamental Theorem of Algebra

The set of real numbers was not sufficient to solve the simple quadratic equation $x^2+1=0$. So we added the zero i of this equation to the real numbers, forming the set of complex numbers. It turns out that the set of complex numbers are **algebraically closed**; that is, every polynomial equation has a (real or) complex solution. The consequences of this fact are so important that we call this the Fundamental Theorem of Algebra.

> **Theorem 4.2.1 (Fundamental Theorem of Algebra).** *Every non-constant polynomial has at least one zero.*

This result was first stated as a theorem by d'Alembert in 1746 (and d'Alembert offered a partial proof). However, the empirical evidence for this theorem was so strong that it was generally accepted as fact long before it was finally stated as a theorem. The first complete proof was given the following century by great mathematician Carl F. Gauss, who published many alternate proofs before his death in 1855. Even today mathematicians continue to discover and publish new proofs, but none will fit easily into this chapter. If you are interested, we sketch Gauss' fourth proof in an appendix.

Let $P(x)$ be a polynomial of degree $n > 0$. By the Fundamental Theorem we know $P(x)$ has a zero r_1, hence the factor $(x - r_1)$. By the Fundamental Theorem again $P(x)/(x - r_1)$ has a zero r_2, hence the factor $(x - r_2)$. Continuing this way we find the following alternate statement of the Fundamental Theorem.

> **Corollary 4.2.2.** *Every polynomial of degree $n > 0$ may be written as the product of n linear factors, that is,*
> $$P(x) = a_0(x - r_1)(x - r_2)\ldots(x - r_n).$$

Example 4.2.1. *$P(x) = 2x^3 - 14x + 12$ has the zeros 1, 2 and -3. So by the Fundamental Theorem of Algebra these are the only zeros of $P(x)$, and by Corollary 4.2.2:*
$$P(x) = 2(x-1)(x-2)(x+3).$$

Example 4.2.2. *A polynomial $P(x)$ has only the zeros 0, 1 and -1. What can we say about the degree of $P(x)$?*

Solution: We can say that the degree of $P(x)$ is at least three. Notice that the degree can be much larger, for example
$$x^{200}(x-1)^{300}(x+1)^{400}$$

4.2. The Fundamental Theorem of Algebra

has degree 900, but still has only the three distinct zeros 0, 1 and -1. ∎

This last example shows that the linear factors of a polynomial P may be equal, that is, a polynomial P need not have n *distinct* zeros. If $(x-r)^m$ $(m>0)$ divides $P(x)$ but $(x-r)^{m+1}$ does not divide $P(x)$, then we say the zero r has **multiplicity** m. If a zero r has multiplicity one, it is called a **simple zero** and is not a multiple zero. If a zero r has multiplicity two (or three) it is called a **double zero** (or **triple zero**). When we count (or list) the zeros of a polynomial we count (list) the zeros of multiplicity m, m times. Using this convention we have a third way to state the Fundamental Theorem of Algebra.

Corollary 4.2.3. *Every polynomial of degree n has exactly n zeros.*

Exercises 4.2:

1. Give an example of a polynomial with exactly 5 zeros, but only 2 distinct zeros.

2. Find a polynomial with the zeros 1, -1, 2 with multiplicities 3, 3, 2 respectively.

3. Find infinitely many quintic polynomials with the zeros 2, -2 and multiplicities 2, 3 respectively.

4. Factor the following polynomials into linear factors:

   a) $x^3 - 1$
   b) $x^4 - 1$
   c) $x^2 + 1$
   d) $x^4 + 1$
   e) $x^2 + x + 1$
   f) $x^4 + x^2 + 1$
   g) $x^2 + i$
   h) $x^4 + x^2 + i$
   i) $x^n - r^n$
   j) $ax^2 + bx + c$
   k) $ax^{2n} + bx^n + c$

5. Factor the polynomial $x^4 + 4$ into the product of two quadratic polynomials with integer coefficients.

6. The Fundamental Theorem of Algebra states that every polynomial of positive degree has at least one zero. Show by example that the zero does not have to be real.

7. (*) Write a short essay on the history of the Fundamental Theorem of Algebra. See almost any book on the History of Mathematics. Gauss gave the first rigorous proof of this theorem in his thesis (1799), so you might also look under Gauss.

8. Prove Corollary 4.2.3.

9. Prove the following restatement of Corollary 4.2.3.

Corollary 4.2.4. *Let $P(x)$ have the g distinct zeros r_i with multiplicities k_i. Then the degree of $P(x)$ is $k_1 + k_2 + k_3 + \ldots + k_g$.*

4. ZEROS OF POLYNOMIALS

10. Prove the following lemma.

    Lemma 4.2.5. *Let r be a zero of $P(x)$ with multiplicity m. If m is odd, then the graph of $y = P(x)$ crosses the x-axis at $x = r$. If m is even then the graph of $y = P(x)$ is tangent to, but does not cross, the x-axis at $x = r$.*

11. (*) Let $r_1, \ldots, r_n$ be the n zeros of the polynomial $P(x)$. Prove the following identity:
    $$P'(x) = \frac{P(x)}{x - r_1} + \frac{P(x)}{x - r_2} + \frac{P(x)}{x - r_3} + \cdots + \frac{P(x)}{x - r_n}.$$

    Hint: Write $P(x) = a_0(x - r_1)(x - r_2)\ldots(x - r_n)$ and use the product rule for derivatives.

12. (*) Writing $\zeta = \cos\frac{2\pi}{n} + i\sin\frac{2\pi}{n}$, show that
    $$x^{n-1} + x^{n-2} + \ldots + x + 1 = (x - \zeta)(x - \zeta^2)\ldots(x - \zeta^{n-1})$$

    Hint: Multiply by $x - 1$.

13. (**) Show that
    $$2^{n-1}\sin\left(\frac{\pi}{n}\right)\sin\left(\frac{2\pi}{n}\right)\sin\left(\frac{3\pi}{n}\right)\cdots\sin\left(\frac{(n-1)\pi}{n}\right) = n.$$

    Hint: Set $x = 1$ in the previous problem and take the absolute value of both sides using
    $$1 - \cos\frac{2m\pi}{n} = 2\sin^2\frac{m\pi}{n}.$$

14. (A) A field R for which every polynomial in $R[x]$ has a zero in R is called an **algebraically closed field**. Prove that if R is a finite field then R is not algebraically closed. Hint: Count.

15. (*) A function is called **analytic** if it has a derivative at every point for which it is defined. Prove the Fundamental Theorem of Algebra using the following.

    Theorem 4.2.6 (Liouville's Theorem). *A function which is both analytic and bounded on the whole complex plane must be a constant.*

    Hint: Assume $P(x)$ has no zeros and consider $1/P(x)$.

16. (*) Let the polynomial P of degree n have the zeros $r_1, \ldots, r_n$ and leading coefficient a. Prove that
    $$P(x) = a(x - r_1)(x - r_2)\ldots(x - r_n)$$

17. (*) The Fundamental Theorem of Arithmetic states that every positive integer can be factored uniquely into the product of distinct primes. How does this compare with Corollary 4.2.3?

4.3 Imaginary Zeros

Recall that the complex conjugate of the number $z = a + bi$ is $a - bi$. We showed that the conjugate of a product is the product of the conjugates, and the conjugate of a sum is the sum of the conjugates. Using these facts and induction we can show:

Lemma 4.3.1. *Let $P(x)$ be a polynomial with real coefficients. If $P(a+bi) = A+Bi$ with a, b, A and B real, then $P(a - bi) = A - Bi$.*

A very useful special case of this result is the following.

Theorem 4.3.2 (Conjugate Zeros). *Let $P(x)$ be a polynomial with real coefficients. If $a + bi$ is a zero of $P(x)$ with a, b real, then $a - bi$ is also a zero.*

Example 4.3.1. *Find a polynomial $P(x)$ with real coefficients and (just) the simple zeros $2 - i$, $4 + i$.*

Solution: $P(x)$ must also have the zeros $2 + i$ and $4 - i$; so by the Extended Factor Theorem we can take
$$\begin{aligned} P(x) &= (x - 2 + i)(x - 2 - i)(x - 4 - i)(x - 4 + i) \\ &= (x^2 - 4x + 5)(x^2 - 8x + 17) \\ &= x^4 - 12x^3 + 54x^2 - 108x + 85. \end{aligned}$$
If the problem had not stated that the zeros were simple, then the answer could have been
$$(x - 2 + i)^n (x - 2 - i)^n (x - 4 - i)^m (x - 4 + i)^m$$
for any positive integers n and m. ∎

Corollary 4.3.3. *Let $P(x)$ be a polynomial with real coefficients. Then $P(x)$ may be factored into the product of linear and quadratic polynomials each of which has real coefficients, and the quadratic factors do not have real zeros.*

4. Zeros of Polynomials

Example 4.3.2. *Solve the polynomial equation*

$$P(x) = x^6 - 2x^5 + 16x^4 - 12x^3 + 85x^2 - 50x + 250 = 0,$$

given that it has the zero $1 + 3i$.

Solution: $P(x)$ must also have the zero $1 - 3i$, so $P(x)$ is divisible by

$$x^2 - 2x + 10 = (x - 1 - 3i)(x - 1 + 3i).$$

We choose to perform this division by using synthetic division twice.

$1+3i$	1	-2	16	-12	85	-50	250
		$1+3i$	-10	$6+18i$	-60	$25+75i$	-250
$1-3i$	1	$-1+3i$	6	$-6+18i$	25	$-25+75i$	0
	1	$1-3i$	0	$6-18i$	0	$25-75i$	
	1	0	6	0	25	0	

Thus $P(x) = (x^2 - 2x + 10)(x^4 + 6x^2 + 25)$.

To factor the second term, $x^4 + 6x^2 + 25$, view it as a quadratic by letting $x^2 = z$. This give us the quadratic $z^2 + 6z + 25$ which has the zeros $-3 \pm 4i$ (you can check this using the quadratic formula). Since $-3 \pm 4i = (1 \pm 2i)^2$, the zeros of $x^4 + 6x^2 + 25$ are $\pm(1 \pm 2i)$.

Putting this all together, we have shown the six zeros of $P(x)$ are

$$1+3i, \quad 1-3i, \quad 1+2i, \quad 1-2i, \quad -1+2i, \quad -1-2i,$$

and $P(x)$ may be written

$$P(x) = (x^2 - 2x + 10)(x^2 - 2x + 5)(x^2 + 2x + 5)$$

or

$$P(x) = (x - 1 - 3i)(x - 1 + 3i)(x - 1 - 2i)(x - 1 + 2i)(x + 1 - 2i)(x + 1 + 2i)$$

depending on what we would like to do next. ∎

Exercises 4.3:

1. Give examples to show that the lemma, theorem and the corollary of this section would be false without the condition "Let $P(x)$ be a polynomial with real coefficients."

2. For each set of zeros below, find a polynomial with real coefficients and those values as zeros. Express your answer in standard form (that is, "multiply it out").

4.3. Imaginary Zeros

  a) $4, 2+3i$

  b) $0, 1+2i, 1+2i$ (a double zero)

  c) $2-\sqrt{-3}, \sqrt{3}-i, 2$

  d) $2+3i, (1+i)^2$

  e) $i, (1+i)/\sqrt{2}$

3. Find all of the zeros of the given polynomials. In each case, one or more zeros is given as a hint.

  a) $x^4 - 4x^3 + 16x^2 - 12x + 39;\ 2+3i$

  b) $x^6 + 4x^4 + 4x^2;\ \sqrt{-2}$ (double zero)

  c) $x^4 + 4;\ 1-i$

  d) $x^6 - x^5 + 2x^3 + 5x^2 - 9x + 18;\ \sqrt{2}-i, -\sqrt{2}+i$

  e) $x^4 + (e+\pi)x^3 + (e\pi+1)x^2 + (e+\pi)x + e\pi;\ i$

4. Let $P(x)$ be a polynomial with real coefficients. Show that if $a+bi$ is a zero of $P(x)$ with multiplicity m, then $a-bi$ is also a zero of multiplicity m.

5. Prove Lemma 4.3.1.

6. Prove Theorem 4.3.2.

7. Prove Corollary 4.3.3. Hint: If $x+(a+bi)$ divides $P(x)$ so does $x-(a+bi)$. The product of these two linear polynomials is $(x-a)^2 + b^2$.

8. Prove that a polynomial with real coefficients and odd degree has at least one real zero.

9. (**R) Show that the theorems of this section are still true if "real" is replaced by rational and "i" by $\theta = \sqrt{5}$. Hint: The results of this section are a direct consequence of the fact that complex conjugation is an automorphism of the complex numbers $\mathbb{C}$ which is the identity on $\mathbb{R}$. So to do this problem, show that the map that takes $a+b\theta$ to $a-b\theta$ is an automorphism of the field $\{a+b\theta \mid a,b \in \mathbb{Q}\}$ which is the identity on $\mathbb{Q}$.

10. (***R) For which integers n (positive, negative or zero), may 5 be replaced by n in the previous problem?

4. Zeros of Polynomials

4.4 Finding Multiple Zeros

Recall r is a zero of $P(x)$ with multiplicity m if and only if $P(x) = (x - r)^m Q(x)$ where $Q(x)$ is a polynomial such that $Q(r) \neq 0$. If a zero r has multiplicity one, it is called a **simple zero** and is not a multiple zero. If a zero r has multiplicity two (or three) it is called a **double zero** (or **triple zero**).

Theorem 4.4.1. *Let $P'(x)$ denote the derivative of $P(x)$ with respect to x. If r is a zero of $P(x)$ with multiplicity $m > 1$, then r is a zero of $P'(x)$ with multiplicity $m - 1$.*

Theorem 4.4.2. *Let $Q(x)$ be the greatest common divisor of $P(x)$ and $P'(x)$. If r is a zero of $P(x)$ with multiplicity $m > 1$, then r is a zero of $Q(x)$ with multiplicity $m - 1$.*

Example 4.4.1. *Solve the following polynomial which has multiple zeros.*

$$P(x) = x^4 + 4x^3 + 8x^2 + 8x + 3$$
$$P'(x) = 4x^3 + 12x^2 + 16x + 8$$

Solution: We begin by finding $Q(x)$, the greatest common divisor of $P(x)$ and $P'(x)$, because any multiple zero of $P(x)$ will be a zero of $Q(x)$. We use the Euclidean algorithm, and start by dividing $P(x)$ by $P'(x)/4$ (why not by $P'(x)$?)

$$\begin{array}{r}
x + 1 \\
x^3 + 3x^2 + 4x + 2 \overline{) x^4 + 4x^3 + 8x^2 + 8x + 3} \\
-x^4 - 3x^3 - 4x^2 - 2x \\
\hline
x^3 + 4x^2 + 6x + 3 \\
-x^3 - 3x^2 - 4x - 2 \\
\hline
x^2 + 2x + 1
\end{array}$$

Next we divide $x^3 + 3x^2 + 4x + 2$ by $x^2 + 2x + 1$.

$$\begin{array}{r}
x + 1 \\
x^2 + 2x + 1 \overline{) x^3 + 3x^2 + 4x + 2} \\
-x^3 - 2x^2 - x \\
\hline
x^2 + 3x + 2 \\
-x^2 - 2x - 1 \\
\hline
x + 1
\end{array}$$

Finally, we see that $x+1$ does divide x^2+2x+1, so $x+1$ is the greatest common divisor of $P(x)$ and $P'(x)$. In follows that -1 is a zero of $P(x)$ with multiplicity two. We divide $P(x)$ by $(x+1)^2$ to find the other zeros of $P(x)$ (we use synthetic division twice):

$$
\begin{array}{r}
x^2+2x+3 \\
x^2+2x+1 \overline{\smash{)}\, x^4+4x^3+8x^2+8x+3} \\
\underline{-x^4-2x^3-x^2} \\
2x^3+7x^2+8x \\
\underline{-2x^3-4x^2-2x} \\
3x^2+6x+3 \\
\underline{-3x^2-6x-3} \\
0
\end{array}
$$

Thus $P(x) = (x+1)^2(x^2+2x+3)$. The zeros of x^2+2x+3 are $-1 \pm 2i$ so the zeros of $P(x)$ are

$$1, \quad 1, \quad -1+2i, \quad -1-2i.$$

■

Exercises 4.4:

1. Solve the following, each of which has multiple zeros:

   a) $x^5 + 2x^4 - 20x^3 - 8x^2 + 128x - 128$
   b) $x^4 - 4x^3 + 8x^2 - 8x + 4$
   c) $x^4 - 6x^3 + 14x^2 - 16x + 8$
   d) $x^5 - x^4 + 2x^3 - 2x^2 + x - 1$
   e) $x^4 - 4x + 3$
   f) $x^3 + 3x^2 - 9x - 27$
   g) $3\sqrt{3}x^3 - 3\sqrt{3}x - 2$
   h) $x^4 + 4x^3 + 2x^2 - 4x + 1$

2. Prove Theorem 4.4.1. Hint: Let $P(x) = (x-r)^m Q(x)$. By the product rule for derivatives we have

   $$P'(x) = (x-r)^{m-1}(mQ(x) + (x-r)Q'(x)).$$

3. Prove the converse of Theorem 4.4.1 is false. (Give an example of a polynomial P for which P' has a zero r which is not a zero of P.)

4. Prove Theorem 4.4.2 using Theorem 4.4.1.

5. Let $P^{(n)}(x)$ denote the nth derivative of P with respect to x. Prove that if r is a zero of P with multiplicity m then r is a simple zero of $P^{(m-1)}(x)$.

6. Prove the following.

4. Zeros of Polynomials

Lemma 4.4.3. *If $P(r) = P'(r) = \ldots = P^{(m-1)}(r) = 0$ but $P^{(m)}(r) \neq 0$, then r is a zero of multiplicity m.*

7. (*) Prove the converse of Theorem 4.4.2:

Lemma 4.4.4. *Let r be a zero of $\gcd(P(x), P'(x))$ with multiplicity $m - 1 > 0$. Then r is a zero of $P(x)$ of multiplicity m.*

4.5 Lagrange's Interpolation Formula

We have seen how to construct a polynomial with given zeros $r_1, r_2, \ldots, r_m$; that is, we can construct a polynomial $P(x)$ that takes on the value 0 at each of these points. It seems natural then to ask if we can find a polynomial $L(x)$ that takes on the values y_i at $x = r_i (i = 1, 2, \ldots, m)$. The positive answer is given in the following.

Theorem 4.5.1 (Lagrange's Interpolation Formula). *Suppose $r_1, r_2, \ldots, r_m$ are distinct numbers and that $y_1, \ldots, y_m$ are any m numbers. Then there is a unique polynomial $L(x)$ of degree less than or equal to $m - 1$ which takes on the values y_i at $x = r_i$; namely*

$$L(x) = y_1 P_1(x) + y_2 P_2(x) + \ldots + y_m P_m(x)$$

where the P_i are defined as follows:

$$P_i = \frac{(x - r_1) \ldots (x - r_{i-1})(x - r_{i+1}) \ldots (x - r_m)}{(r_i - r_1) \ldots (r_i - r_{i-1})(r_i - r_{i+1}) \ldots (r_i - r_m)}.$$

$L(x)$ is called the **Lagrange Interpolating Polynomial**.

Example 4.5.1. *Find the Lagrange Interpolating Polynomial defined by the following points on the Cartesian plane: $(0, 1)$, $(1, 2)$, $(2, 4)$, and $(3, 8)$.*

4.5. Lagrange's Interpolation Formula

Solution: Using the definitions of $P_i(x)$ and r_i we find the following.

$$P_1(x) = \frac{(x-1)(x-2)(x-3)}{(0-1)(0-2)(0-3)} = -\frac{1}{6}(x^3 - 6x^2 + 11x - 6)$$

$$P_2(x) = \frac{(x-0)(x-2)(x-3)}{(1-0)(1-2)(1-3)} = \frac{1}{2}(x^3 - 5x^2 + 6x)$$

$$P_3(x) = \frac{(x-0)(x-1)(x-3)}{(2-0)(2-1)(2-3)} = -\frac{1}{2}(x^3 - 4x^2 + 3x)$$

$$P_4(x) = \frac{(x-0)(x-1)(x-2)}{(3-0)(3-1)(3-2)} = \frac{1}{6}(x^3 - 3x^2 + 2x)$$

Thus we find

$$L(x) = 1P_1(x) + 2P_2(x) + 4P_3(x) + 8P_4(x) = \tfrac{1}{6}x^3 + \tfrac{5}{6}x + 1.$$

■

Exercises 4.5:

1. Show that the Lagrange Interpolating Polynomial is given by

    a) $m = 1$: $L(x) = y_1$

    b) $m = 2$: $L(x) = (y_1(x - r_2) - y_2(x - r_1))/(r_2 - r_1)$

    c) $m = 3$: $L(x) = y_1 \frac{(x-r_2)(x-r_3)}{(r_1-r_2)(r_1-r_3)} + y_2 \frac{(x-r_1)(x-r_3)}{(r_2-r_1)(r_2-r_3)} + y_3 \frac{(x-r_2)(x-r_1)}{(r_3-r_2)(r_3-r_2)}$

2. (a) Show by example that $L(x)$ may have degree less than $m - 1$. (b) Under what conditions is the degree of $L(x)$ equal to $m - 1$?

3. Prove the Lagrange Interpolation Formula. Hint: Notice $P_i(r_j) = 1$ if $i = j$, and 0 if $i \neq j$. Use this to show that $L(x)$ takes on the values as advertised. Then prove any other polynomial taking on these values equals $L(x)$ by using the Identity Theorem (see page 53).

4. Find the Lagrange Interpolating Polynomials through each of the following sets of points.

    a) (1,0), (2,1), (4,2), (8,3)

    b) (1,0), (2,1), (3,0), (4,1), (5,0)

    c) (−1,1), (0,−1), (1,1)

    d) (0,0), (1,100), (2,−100), (3,0)

5. Often you encounter problems such as "What is the next number in the sequence 1, 1, 2, 3, 5, 8, ...?" Use Lagrange's Interpolating Polynomial to show that given terms $a_1, a_2, a_3, \ldots, a_n$, the next term a_{n+1} may be any number you choose (real or complex).

4. Zeros of Polynomials

6. Graph the points and interpolating polynomial from Example 4.5.1 along with the graph of $y = 2^x$ (which goes through the same four points). State which function better captures the trend of the points—**explain what 'better' means** in your answer.

7. (*) Let $P(x)$ be a polynomial of degree nine for which
$$P(1) = \frac{1}{1}, \quad P(2) = \frac{1}{2}, \quad P(3) = \frac{1}{3}, \quad \ldots, \quad P(10) = \frac{1}{10}.$$
What is $P(11)$? Hint: Possible method one: Lagrange Interpolating Polynomial—direct but cumbersome and boring. Possible method two: Let $G(x) = xP(x) - 1$. What is $G(1)$, $G(2)$, $G(3)$, ..., $G(10)$? What is $G(0)$? These answers are enough to determine $G(11)$, and then $P(11)$.

8. Write a program which calculates the coefficients of $L(x)$ when given an integer n and the n points (r_i, y_i).

9. (*) With the Lagrange Interpolating Polynomial we can determine an nth degree polynomial $P(x)$ if we have $n+1$ distinct points on $y = P(x)$, but at times we need far fewer points! Suppose $P(x)$ is a polynomial with non-negative integer coefficients.

    a) If $P(1) = 10$ and $P(10) = 32014$, what is $P(x)$?
    b) If $P(1) = 31$ and $P(31) = 57414001$, what is $P(x)$?

10. (*) Expand on the previous problem by showing the following.

Lemma 4.5.2. *Suppose $P(x)$ is a polynomial with non-negative integer coefficients. We may determine $P(x)$ from $P(1)$ and $P(P(1))$.*

11. Let $P(x)$ and $Q(x)$ be polynomials with $\deg(P) < \deg(Q)$. If Q has distinct zeros $r_1, r_2, \ldots, r_n$, we may write
$$\frac{P(x)}{Q(x)} = \sum_{i=1}^{n} \frac{R_i}{x - r_i}. \tag{4.1}$$
(This is called the partial fraction decomposition in calculus.) Prove that
$$R_i = \frac{P(r_i)}{Q'(r_i)}.$$
Hint: Multiply Equation 4.1 by $x - r_i$, differentiate, and then evaluate at appropriate values.

12. (**) Given $(x_1, y_1), \ldots, (x_n, y_n)$ with the x_i distinct, define
$$Q(x) = (x - x_1)(x - x_2) \cdots (x - x_n),$$
and $R_i = \frac{y_i}{Q'(x_i)}$. Finally write $\sum_{i=1}^{n} \frac{R_i}{x - x_i}$ as $\frac{P(x)}{Q(x)}$. Show that $P(x)$ is the Lagrange Interpolating Polynomial.

Elementary Techniques for Finding Zeros

The imaginary numbers are a wonderful flight of God's Spirit; they are almost an amphibian between being and not being.

G. W. Leibniz, 1702

Reeling and Writhing, of course, to begin with, the Mock Turtle replied And the different branches of Arithmetic—Ambition, Distraction, Uglification, and Derision

Lewis Carroll, Alice in Wonderland

Chapter 5

Elementary Techniques for Finding Zeros

In this chapter we continue our study of the zeros of polynomials, presenting methods for finding zeros that are either integers or rational numbers. We first present several methods for finding rational zeros or showing that rational zeros do not exist. We next study reciprocal polynomials, a type of polynomial which is especially easy to solve. In the final section we discuss Lill's amusing method of finding real zeros—a method that acts like an odd combination of billiards and geometry.

5.1 Rational Zeros

A NUMBER THAT CAN BE EXPRESSED as the quotient of two integers is said to be **rational**. Real numbers that cannot be expressed as the quotient of two integers are said to be **irrational**.

Theorem 5.1.1 (Rational Zero Theorem). *If the equation with integral coefficients*
$$a_0 x^n + a_1 x^{n-1} + \ldots + a_n = 0$$
has a rational zero r/s, where r and s are relatively prime integers, then r is a divisor of a_n and s is a divisor of a_0.

Proof. In the equality in the theorem, replace x by the zero r/s, multiply by s^n, then subtract $a_n s^n$ to get
$$a_0 r^n + a_1 r^{n-1} s + a_2 r^{n-2} s^2 + \ldots + a_{n-1} r s^{n-1} = -a_n s^n.$$

5. Elementary Techniques for Finding Zeros

Note r divides the left hand side so it also divides the right hand side. The number r does not divide s^n (because r and s have no factor in common) so r divides a_n. Similarly we see s divides a_0. □

Example 5.1.1. *Find all rational zeros of*
$$4x^4 - 11x^2 + 9x - 2.$$

Solution: By the Rational Zero Theorem we have the following possibilities for the rational zeros:
$$2, -2, 1, -1, \frac{1}{2}, -\frac{1}{2}, \frac{1}{4}, -\frac{1}{4}.$$
Using synthetic division we quickly see that the zeros are -2, 1, $\frac{1}{2}$, and $\frac{1}{2}$. ■

Example 5.1.2. *Find all the rational zeros of*
$$12x^4 + 36x^3 - x^2 - 84x - 63.$$

Solution: If we have a graphing device (perhaps a calculator) we can speed up our work considerably. By the Rational Zero Theorem the possible rational zeros are plus or minus each of the following:
$$1, 3, 7, 9, 21, 63, \frac{1}{2}, \frac{3}{2}, \frac{7}{2}, \frac{9}{2}, \frac{21}{2}, \frac{63}{2}, \frac{1}{3}, \frac{7}{3}, \frac{1}{4}, \frac{3}{4}, \frac{7}{4}, \frac{9}{4}, \frac{21}{4}, \frac{63}{4}, \frac{1}{6}, \frac{7}{6}, \frac{1}{12}, \frac{7}{12}.$$

A quick graph shows that most of the zeros appear close to the origin (Figure 5.1, left).

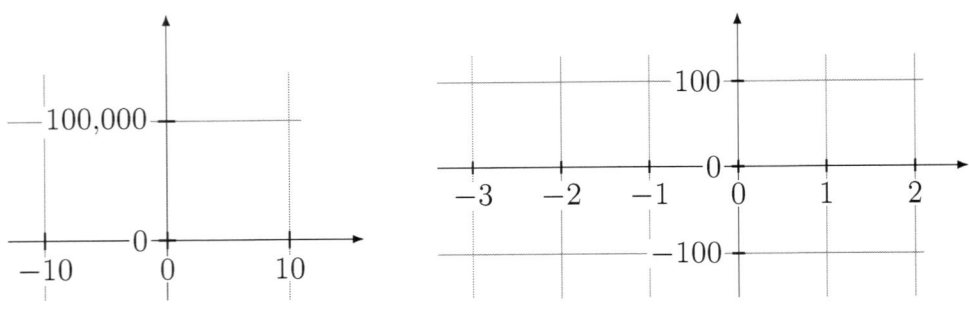

Figure 5.1: Graphs of $12x^4 + 36x^3 - x^2 - 84x - 63$

A closer look show there is at least one zero between -2 and -1 (Figure 5.1, right). This means we should test $-\frac{3}{2}, -\frac{7}{3}, -\frac{3}{4}$, and $-\frac{7}{4}$. Actually, there are three zeros between -2 and -1, but that is not easy to see from these graphs, even when we look much closer (Figure 5.2). Using synthetic division we find $-\frac{3}{2}$ is a double zero and leaves the quotient $3x^2 - 7$, so this polynomial does not have any further rational zeros. ■

5.1. Rational Zeros

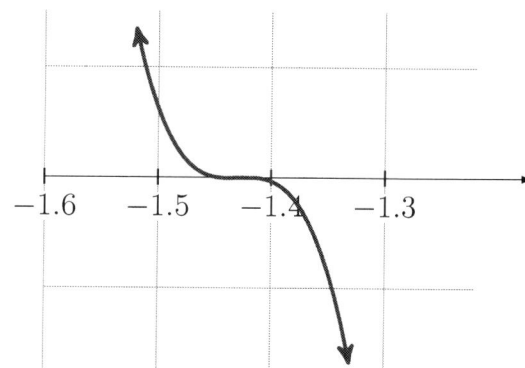

Figure 5.2: Graph of $12x^4 + 36x^3 - x^2 - 84x - 63$ with $-0.03 \leq y \leq 0.03$

This example showed that the use of graphs can be very helpful, but in practice students often have great difficulty in determining appropriate windows in which to view the graphs (so they often just use the default window and end up wondering why they do not see anything). In Chapter 9 we discuss ways we can determine where the zeros might be found in order to help choose a good graphing window.

Example 5.1.3. *Show that the square root of two is irrational.*

Solution: We can do this by showing that $x^2 - 2 = 0$ has no rational zeros. By the Rational Zero Theorem, if this equation has a rational zero, it must be among $1, -1, 2, -2$. A quick check shows that none of these are zeros, so the square root of two is irrational. ∎

Exercises 5.1:

1. First list the possible rational zeros, then find the rational zeros of the following polynomials.

    a) $x^3 + 3x^2 - 4x - 12$
    b) $x^5 + 3x^4 - 5x^3 - 15x^2 + 4x + 12$
    c) $6x^3 - 19x^2 + 19x - 6$
    d) $4x^5 + 16x^4 + 9x^3 - 35x^2 - 51x - 18$
    e) $3x^4 + 5x^3 - 29x^2 - 45x + 18$
    f) $18x^4 + 19x^3 - 3x - 2$
    g) $15x^4 - x^3 - 13x^2 - x - 28$
    h) $21x^4 - x^3 + 11x^2 - x - 10$
    i) $18x^5 - x^4 - 19x^3 + x^2 - x + 2$

5. Elementary Techniques for Finding Zeros

j) $x^4 + \frac{19}{18}x^3 - \frac{1}{6}x - \frac{1}{9}$

k) $\frac{15}{28}x^4 - \frac{1}{28}x^3 - \frac{13}{28}x^2 - \frac{1}{28}x - 1$

2. Use the Rational Zero Theorem to prove that the square root of six is irrational.

3. Use the Rational Zero Theorem to prove that the golden ratio is irrational. (The golden ratio is the largest zero of $x^2 + x + 1 = 0$.)

4. (**) Prove the following.

> **Lemma 5.1.2.** *Let P be a polynomial with integral coefficients. If r and s are relatively prime integers such that r/s is a zero of P, then $P(x) = (sx - r)Q(x)$ where $Q(x)$ has integral coefficients.*

Hint: Induction. If $Q(x) = a_0 x^n + \ldots + a_n$, then r/s is also a zero of

$$(a_0 r/s + a_1)x^{n-1} + a_2 x^{n-2} + \ldots + a_0.$$

Note that $a_0 r/s$ is an integer.

5. (*) Prove the following.

> **Theorem 5.1.3.** *Let P be a polynomial with integral coefficients. If r and s are relatively prime integers such that r/s is a zero of P, then $r - s$ divides $P(1)$ and $r + s$ divides $P(-1)$.*

Hint: Use Lemma 5.1.2.

6. Show how the theorem in the previous problem may be used to reduce the number of possible rational zeros. Hint: What are the possible zeros of (e) in problem one? Notice that when checking P for rational zeros we must usually calculate $P(\pm 1)$ anyway—why?

7. (N) Let n and m be integers. Show that if n is not a perfect m^{th} power, then $n^{1/m}$ is irrational.

8. (*) Show that $\sqrt{2} + \sqrt{3}$ is irrational. Hint: Set $x = \sqrt{2} + \sqrt{3}$, square, rearrange and square again to find a polynomial with integral coefficients which x satisfies.

9. (*) Show that if a polynomial of degree less than or equal to five with rational coefficients has multiple zeros then it also has a rational zero, except in the case when the degree is four and the polynomial a perfect square.

5.2 Integral Zeros of Polynomials

If a zero of a polynomial is an integer, then it is said to be an **integral zero** of the polynomial. From the Rational Zero Theorem we have the following.

Theorem 5.2.1 (Integral Zero Theorem). *Let $P(x)$ be a polynomial with integral coefficients. If r is an integral zero of $P(x)$, then r divides $a_n = P(0)$.*

Theorem 5.2.2. *Let $P(x)$ be a polynomial with integral coefficients. If the leading coefficient of $P(x)$ is 1 then any rational zero of $P(x)$ is an integral zero.*

This theorem may also be worded: If $P(x)$ is a monic polynomial with integral coefficients, then each zero of $P(x)$ is either an integer or irrational.

Example 5.2.1. *Find the rational zeros of*
$$x^5 + x^4 - 20x^3 - 44x^2 - 21x - 45.$$

Solution: By the theorems we know any rational zero is an integral zero. The possible zeros are:

$$1, -1, 3, -3, 5, -5, 9, -9, 15, -15, 45, -45.$$

By synthetic division we find $1, -1$ and 3 are not zeros. Next we try -3:

$$\begin{array}{r|rrrrrr}
-3 & 1 & 1 & -20 & -44 & -21 & -45 \\
   &   & -3 & 6 & 42 & 6 & 45 \\
\hline
   & 1 & -2 & -14 & -2 & -15 & 0
\end{array}$$

We have to try -3 again to see if it was a multiple zero.

$$\begin{array}{r|rrrrr}
-3 & 1 & -2 & -14 & -2 & -15 \\
   &   & -3 & 15 & -3 & 15 \\
\hline
   & 1 & -5 & 1 & -5 & 0
\end{array}$$

$$\begin{array}{r|rrrr}
-3 & 1 & -5 & 1 & -5 \\
   &   & -3 & 24 & -75 \\
\hline
   & 1 & -8 & 25 & -80
\end{array}$$

5. ELEMENTARY TECHNIQUES FOR FINDING ZEROS

So we see -3 was a double, but not a triple zero. (Notice that by the Integral Zero Theorem this last division was not necessary, because 3 does not divide -5.) Now the only possible zeros are 5 and -5; we will first try 5.

$$\begin{array}{r|rrrr} 5 & 1 & -5 & 1 & -5 \\ & & 5 & 0 & 5 \\ \hline & 1 & 0 & 1 & 0 \end{array}$$

Thus the equation has the zeros $-3, -3, 5, i,$ and $-i$. ∎

Example 5.2.2. *Find the rational zeros of*

$$x^5 + 2x^4 - 34x^3 + 28x^2 + 129x - 126.$$

Solution: The integer 126 has the prime factorization $2 \cdot 3 \cdot 3 \cdot 7$, so the possible rational zeros are as follows.

$$1, -1, 2, -2, 3, -3, 6, -6, 7, -7, 9, -9, 14, -14,$$
$$18, -18, 21, -21, 42, -42, 63, -63, 126, -126.$$

We easily see that 1 is a zero (by just adding the coefficients in our head), so we begin by synthetically dividing by 1.

$$\begin{array}{r|rrrrrr} 1 & 1 & 2 & -34 & 28 & 129 & -126 \\ & & 1 & 3 & -31 & -3 & 126 \\ \hline & 1 & 3 & -31 & -3 & 126 & 0 \end{array}$$

$$\begin{array}{r|rrrrr} 1 & 1 & 3 & -31 & -3 & 126 \\ & & 1 & 4 & -27 & -30 \\ \hline & 1 & 4 & -27 & -30 & 96 \end{array}$$

Using synthetic division we see that 2 is not a zero, so we next check -2.

$$\begin{array}{r|rrrrr} -2 & 1 & 3 & -31 & -3 & 126 \\ & & -2 & -2 & 66 & -126 \\ \hline & 1 & 1 & -33 & 63 & 0 \end{array}$$

Thus -2 is a zero, but not a double zero (as 63 is odd). In fact, the possible zeros now are just

$$3, -3, 7, -7, 9, -9, 21, -21, 63, -63.$$

5.2. Integral Zeros of Polynomials

We next try 3 finding that it is a double zero and leaving the quotient $x + 7$. This shows that the zeros are $1, -2, 3, 3$, and -7. ∎

One way of shortening our work is contained in the following theorem.

Theorem 5.2.3. *Let P have integral coefficients. If r is an integral zero of P and s is any integer, then $r - s$ divides $P(s)$.*

Example 5.2.3. *Show that the following polynomial has no rational zeros.*

$$P(x) = x^4 + x^3 + 11x^2 - 49x + 2940$$

Solution: The constant term 2940 has the prime factorization $2 \cdot 2 \cdot 3 \cdot 5 \cdot 7 \cdot 7$, so we have the following 72 possible rational zeros.

$$1, -1, 2, -2, 3, -3, 4, -4, 5, -5, 6, -6, 7, -7, 10, -10, 12, -12, 14,$$
$$-14, 15, -15, 20, -20, 21, -21, 28, -28, 30, -30, 42, -42, 49, -49,$$
$$60, -60, 70, -70, 84, -84, 98, -98, 105, -105, 140, -140, 147,$$
$$-147, 196, -196, 210, -210, 245, -245, 294, -294, 420, -420, 490,$$
$$-490, 588, -588, 735, -735, 980, -980, 1470, -1470, 2940, -2940$$

We could use synthetic division to check each number in this list, but this would take a long time. First we calculate $P(1) = 2904$ which has the prime factorization $2 \cdot 2 \cdot 2 \cdot 3 \cdot 11 \cdot 11$. So we see 1 is not a zero and by Theorem 5.2.3, if r is a zero, then $r-1$ divides 2904. Checking the list above against this, we find the possible zeros are

$$-1, 2, -2, 3, -3, 4, 5, -5, 7, -7, 12, -21.$$

We have ruled out 60 potential zeros with one function evaluation! Next we find $P(-1) = 3000$, so -1 is not a zero and by Theorem 5.2.3, $r+1$ divides 3000. The possible zeros are now

$$2, -2, 3, -3, 4, 5, -5, 7, -7, -21.$$

Trying the next possibility 2 we find $f(2) = 2910$ (prime factors 2, 3, 5, 97), so $r-2$ divides 2910 and the possible remaining zeros are

$$-2, 3, -3, 4, 5, 7.$$

We now find $P(-2) = 3090$ ruling out $-2, 5$ and 7. Finally, we find that $P(3)$ and $P(4)$ are not zeros ($P(4) = 3240$ rules out -3). This shows $P(x)$ has no rational zeros (and we showed this by trying only six of the 72 possible rational zeros). ∎

5. Elementary Techniques for Finding Zeros

Exercises 5.2:

1. First list the possible rational zeros, then find the rational zeros of the following polynomials (if any).

   a) $x^3 + 3x^2 - 4x - 12$

   b) $x^5 + 3x^4 - 5x^3 - 15x^2 + 4x + 12$

   c) $x^4 - 13x^3 + 8x^2 + 52x - 48$

   d) $x^3 - 18x^2 + 108x - 216$

2. Find the rational zeros of the following polynomials (if any).

   a) $x^3 - 18x^2 + 18x - 21$

   b) $x^4 - 24x^3 + 168x^2 - 576x + 3456$

   c) $x^5 - 13x^4 + 36x^3 - 468x^2 + 288x - 3744$

3. Let $P(x)$ be a polynomial with integer coefficients. Show that if $P(0)$ and $P(P(0))$ are both odd, then $P(x)$ has no integer zeros.

4. Prove the Integral Zero Theorem both by using the Rational Zero Theorem and then directly (in the manner the Rational Zero Theorem was proved).

5. Prove Theorem 5.2.2.

6. Prove Theorem 5.2.3.

7. Use Theorem 5.2.3 to find the rational zeros of the polynomial

   $$P(x) = x^5 + 9x^4 - 31x^3 + 780x^2 + 7920x - 8640.$$

   Hint: The constant term factors easily but has 112 divisors! Since $P(1)$ is a relatively small integer (and not zero) you will see both that 1 is not a zero and that if r is, then $r - 1$ divides $P(1)$. This will rule out over 100 of the 112 possibilities!

8. Let n, k, and r be any integers with n, k positive and r odd. Let $P(x)$ be any polynomial with integral coefficients. Use part a of the following problem to show that (a) $x^n + x^k + r$ and (b) $2P(x) + r$ have no integral zeros.

9. Let P be an integral polynomial. Prove the following:

   a) If 2 does not divide $P(0)P(1)$, then P has no integral zeros.

   b) If 6 does not divide $P(0)P(1)P(2)$, then P has no integral zeros.

   c) If 24 does not divide $P(0)P(1)P(2)P(3)$, then P has no integral zeros.

   Hint: For part (a) assume P does have an integral zero r, so $x - r$ divides P and $(0 - r)(1 - r)$ divides $P(0)P(1)$.

10. (N) Recall n **factorial** (denoted $n!$) is the product of the first n positive integers. Generalize the results in the previous problem by proving the following.

> **Theorem 5.2.4.** Let P be an integral polynomial. If $n!$ does not divide
> $$P(0)P(1)\cdots P(n-1)$$
> then P has no integral zeros.

11. (N) Let k be any integer. Show that in the previous problem the product
$$P(0)P(1)\cdots P(n-1)$$
may be replaced by
$$P(k)P(k+1)\cdots P(k+n-1).$$

12. (N) Prove the following lemma.

> **Lemma 5.2.5.** Let $P(x)$ be an integral polynomial. If there is a positive integer n for which $P(x)$ does not have a zero modulo n, then $P(x)$ does not have any integer zeros.

Hint: Contrapositive?

13. (N) Apply the previous problem to example 5.2.3. Hint: We do not want n to divide the constant term (or 0 is a zero modulo n), so the first reasonable choices of n are 8, 9, 11 and 13. Three of these will work.

5.3 Eisenstein's Criterion

Recall that a positive integer is **composite** if it can be written as the product of two integers, each greater than one. For example, $6 = 2 \cdot 3$, $12 = 2 \cdot 6$ and $51 = 3 \cdot 17$ are composite. An integer greater than one which is not composite is called **prime**. So 2, 3, 5 and 7 are prime. The integer one is neither prime nor composite.

Similarly we say a polynomial with rational coefficients is **reducible** if it can be written as a product of two non-constant polynomials each with rational coefficients. For example, $x^2 - 1 = (x-1)(x+1)$ and $x^3 - 8 = (x-2)(x^2+x+1)$ are reducible. If a polynomial is not reducible, then it is said to be **irreducible**. For instance $x^2 + 1$, $x^3 - 6$, and all linear polynomials are irreducible.

To connect the notion of reducibility with the existence of rational zeros we may prove the following (exercise 1).

5. Elementary Techniques for Finding Zeros

Lemma 5.3.1. *Let $P(x)$ be a polynomial with rational coefficients and degree greater than one. If $P(x)$ has a rational zero, then it is reducible; that is, if $P(x)$ is irreducible, then it does not have a rational zero.*

Notice that the converse is not true; that is, many reducible polynomials do not have rational zeros (exercise 4).

The main result of this section is the following simple, but exceptionally powerful and often used, test for irreducibility.

Theorem 5.3.2 (Eisenstein's Criterion). *Suppose*

$$P(x) = a_0 x^n + a_1 x^{n-1} + a_2 x^{n-2} + \ldots + a_{n-1} x + a_n$$

is a polynomial with integer coefficients and degree at least two. If there is a prime number q such that

1. *q does not divide a_0,*
2. *q divides $a_1, a_2, \ldots, a_n$,*
3. *q^2 does not divide a_n,*

then $P(x)$ is irreducible.

Example 5.3.1. *Show that*

$$P(x) = 2x^9 + 28x^7 - 49x^4 + 21x^3 + 91x - 7$$

has no rational zeros.

Solution: We may use Eisenstein's Criterion with $q = 7$; for 7 divides every coefficient except the first and $7^2 = 49$ does not divide the constant term. (Every positive integer divides 0, so 7 does divide the coefficients of the x^8, x^6, x^5, and x^2 terms.) This shows that the polynomial has no rational zeros, but it also shows that this polynomial cannot be factored into the product of two other polynomials with integer coefficients. ■

Example 5.3.2. *Let k be any integer. Show that $P(x) = x^5 + (25k+4)$ has no rational zeros.*

5.3. Eisenstein's Criterion

Solution: Without knowing k, we do not know the prime factors of $25k+4$, so we cannot use Eisenstein's Criterion directly. However, if we let $Q(x) = P(x+1)$ we will get a polynomial to which we can apply Eisenstein's Criterion.

$$Q(x) = x^5 + 5x^4 + 10x^3 + 10x^2 + 5x + 5(5k+1).$$

The polynomial $Q(x)$ has rational zeros if and only if $P(x)$ does (and $Q(x)$ is irreducible if and only if $P(x)$ is). Notice that 5 divides every coefficient other than the leading coefficient. Also 5 does not divide $5k+1$ so 25 does not divide the constant term. By Eisenstein's Criterion with $p = 5$, $Q(x)$ (hence $P(x)$) is irreducible, so has no rational zeros. ∎

The shift by one in the previous example may seem like magic, and in a way it is, but it is common to shift polynomials this way before applying Eisenstein's Criterion. We do not shift unless we have no choice, and not always by one. For example, to show $P(x) = x^2 + x + 2$ irreducible we might look at $P(x+1) = x^2 + 3x + 4$ (no help), $P(x+2) = x^2 + 5x + 8$ (no help), and then finally $P(x+3) = x^2 + 7x + 14$ (bingo!). After a little trial and error this type of shift gets easier; but do not waste much time shifting; there is not always a shift that will work.

Let's look at another such example, one that will be important to us in later chapters.

Example 5.3.3. *Let p be any prime. Show that the following polynomial is irreducible.*

$$P(x) = x^{p-1} + x^{p-2} + x^{p-3} + \ldots + x^2 + x + 1$$

Solution: No prime divides the constant term 1, so Eisenstein's Criterion does not apply here directly either. Let's try our magic shift from the last example and take a look at $P(x+1)$. Note that $P(x) = \dfrac{x^p - 1}{x - 1}$, so by the Binomial Theorem

$$P(x+1) = \frac{(x+1)^p - 1}{x} = x^{p-1} + px^{p-2} + \frac{p(p-1)}{1 \cdot 2}x^{p-3} + \frac{p(p-1)(p-2)}{1 \cdot 2 \cdot 3}x^{p-4} + \cdots + p.$$

We know p is prime, so it divides every coefficient except the leading one, and p^2 does not divide the constant term, so $P(x+1)$ (and $P(x)$) are irreducible by Eisenstein's Criterion with the prime p. In particular this means that $P(x)$ has no rational zeros. ∎

Important note: Eisenstein's Criterion is a necessary, but not a sufficient condition for irreducibility. That is, there are many polynomials which are irreducible but their irreducibility cannot be shown by Eisenstein' Criterion.

5. Elementary Techniques for Finding Zeros

Exercises 5.3:

1. Prove Lemma 5.3.1. Hint: If $P(x)$ has a rational zero r, then by the Factor Theorem $P(x) = (x - r)Q(x)$. To show that $P(x)$ is reducible, it is necessary to show $Q(x)$ has rational coefficients *and* that $Q(x)$ is non-constant.

2. Show that the following polynomials are irreducible.

    a) $x^4 + 15x^3 + 35x^2 - 60x + 205$
    b) $37x^6 - 26x^5 + 39x^2 - 91x + 273$
    c) $6x^8 + 256x^6 + 356x^3 - 60x + 52$
    d) $x^3 - 2$
    e) $x^3 - 3x + 1$
    f) $2x^7 + 3x^5 - 18x^2 + 6x + 30$
    g) $x^4 + 2x^3 + 4x^2 + 8x + 62$
    h) $x^2 - 7x + 7$
    i) $4x^5 - 14x^4 - 91x^3 + 98x - 21$
    j) $x^6 - 68x^4 - 105x^2 + 51$

    Hints: For (c) divide by out any common division of all of the coefficients. Note (e) requires a shift.

3. Show that $26x^4 + 40x^3 + 356x^2 - 50x + 205$ is irreducible. Hint: Eisenstein's Criterion would have applied if the coefficients were reversed in order, that is, Eisenstein's Criterion applies to the polynomial $x^4 P(1/x) = 205x^4 - 50x^3 + 356x^2 + 40x + 26$. Explain what q to use and why $P(x)$ is irreducible if and only if $x^4 P(1/x)$ is.

4. Prove that the converse of Lemma 5.3.1 is not true, that is, give infinitely many examples of reducible polynomials which do not have any rational zeros. Hint: If they do not have rational zeros, what type of zeros do they have? What is the simplest example you can think of? What if you raise your example to the 100th power. Is it still reducible? Still no rational zeros?

5. Prove that if k is an integer then $x^3 + 3x + (9k + 2)$ is irreducible.

6. Prove that if k is an integer then $x^4 + 2x^2 + (4k - 1)$ is irreducible.

7. Let p be a prime and recall that $n! = n(n-1)(n-2)\ldots(2)(1)$. Show that the following polynomial is irreducible.

$$1 + x + \frac{x^2}{2!} + \frac{x^3}{3!} + \cdots + \frac{x^p}{p!}.$$

Hint: Multiply by $p!$ and use Eisenstein's Criterion.

8. Prove the following.

5.3. Eisenstein's Criterion

Lemma 5.3.3. *Let $P(x)$ be a cubic polynomial with rational coefficients. Then either $P(x)$ has a rational zero or $P(x)$ is irreducible.*

Hint: If it is reducible, what do you know about the degree of its possible factors?

9. (**) Prove Eisenstein's Criterion. Hint: Prove the result by contradiction; that is, let $P(x)$ satisfy the criterion and assume $P(x)$ is reducible ($P(x) = Q(x)R(x)$) to reach a contradiction. Notice the constant term of P is the product of the constant terms of Q and R, so the prime p divides either $Q(0)$ or $R(0)$ (but not both—why?). Suppose it divides $R(0)$. Use this to show p must divide the coefficient of the x term of $R(0)$. Then show it must divide the x^2 term also..., so p divides every coefficient of $R(x)$ and this is a contradiction—why?

10. A polynomial with real coefficients $P(x)$ is **reducible over the reals** if it can be written as the product of two non-constant polynomials each with real coefficients. Prove that every polynomial of degree greater than two is reducible over the reals.

11. (N) Let m be an integer. A polynomial with integral coefficients $P(x)$ is **irreducible modulo m** if and only if $P(x)$ is not the product of two non constant polynomials modulo m, and **reducible modulo m** if it is. Show that x^2+1 is irreducible modulo 3, 4, 6, 7, 8, and 9; but reducible modulo 2, 5, and 10.

12. (N) Notice that Eisenstein's Criterion essentially shows that $P(x)$ is irreducible modulo p^2. Prove the following more general result.

Theorem 5.3.4. *Let $P(x)$ be a polynomial with integral coefficients. If there exists an integer m, relatively prime to the leading coefficient of $P(x)$, such that $P(x)$ is irreducible modulo m, then $P(x)$ is irreducible over the rationals.*

13. (*N) Show x^4+1 is reducible modulo m for all integers m but not reducible over the rationals. Thus the converse of the theorem in the previous problem is false. Hint: First prove that x^4+1 is reducible modulo p for all odd primes p (try a few small primes, find a pattern in the way it factors, use quadratic reciprocity). Second, use induction to show that it is reducible modulo p^i for all i. Third, use the Chinese Remainder Theorem to show this is enough.

14. (*R) Let R be a ring. Define 'reducible' and 'irreducible' over R. Give an example of a ring R for which "Eisenstein's Criterion" holds for R, and example of one for which it does not.

15. (**N) Generalize the theorem in problem 12 to more general rings. Hint: There is a canonical homomorphism from R to the integers modulo p. Generalize this result to any two homomorphic rings.

5. Elementary Techniques for Finding Zeros

5.4 Reciprocal Polynomials

Let $P(x)$ be a polynomial of degree n greater than one, say

$$P(x) = a_0 x^n + a_1 x^{n-1} + a_2 x^{n-2} + \ldots + a_{n-1} x + a_n.$$

Then $P(x)$ is called a **reciprocal polynomial** if one of the following equivalent[1] conditions hold:

1. $P(x) = x^n P(1/x)$.

2. $a_0 = a_n$; and if r is a zero of P, then $1/r$ is also a zero.

3. $a_0 = a_n$; $\quad a_1 = a_{n-1}$; $\quad a_2 = a_{n-2}$; $\quad a_3 = a_{n-3}$; $\quad \ldots$

For example, each of the following are reciprocal polynomials.

- $x^2 + 3x + 1$
- $x^5 + 5x^4 - 3x^3 - 3x^2 + 5x + 1$
- $31x^7 + 2x^4 + 2x^3 + 31$
- $(x^n - 1)/(x - 1) = x^{n-1} + x^{n-2} + \ldots + x^2 + x + 1$.

If P is a reciprocal polynomial, then $P(x) = 0$ is called a **reciprocal equation.** Reciprocal equations are usually easier to solve than other polynomial equations. In fact, to solve a reciprocal polynomial equation of degree $2n$ or $2n + 1$ we need only solve a polynomial equation of degree n, and then n quadratics. Before showing this we need a lemma.

Lemma 5.4.1. *Let m be an integer and set $y = x + x^{-1}$. Then $x^m + x^{-m}$ may be written as a polynomial $Q_m(y)$ of degree m.*

For example,

$$\begin{aligned}
x^0 + x^{-0} &= Q_0(y) &&= 2 \\
x^1 + x^{-1} &= Q_1(y) &&= y \\
x^2 + x^{-2} &= Q_2(y) &&= y^2 - 2 \\
x^3 + x^{-3} &= Q_3(y) &&= y^3 - 3y \\
x^4 + x^{-4} &= Q_4(y) &&= y^4 - y^3 - 3y^2 + 3y
\end{aligned}$$

---

[1] Two conditions are said to be equivalent if whenever one is true, they both are true; and if whenever one is false, they both are false. A set of conditions are called equivalent conditions if each pair are equivalent.

5.4. Reciprocal Polynomials

In fact we may easily show

$$x^{m+1} + x^{-m-1} = Q_{m+1}(y) = yQ_m(y) - Q_{m-1}(y).$$

(This is left to the reader as exercise 1.)

Theorem 5.4.2. *Let $P(x)$ be a reciprocal polynomial of degree $2n$, say*

$$P(x) = a_0 x^{2n} + a_1 x^{2n-1} + \ldots + a_{2n-1}x + a_{2n}.$$

Then the zeros of $P(x)$ are the zeros of the quadratic equations

$$x^2 - y_j x + 1 = 0 \quad (j = 0, 1, \ldots, n)$$

where y_j are the zeros of the following polynomial of degree n

$$a_0 Q_n(y) + a_1 Q_{n-1}(y) + \ldots + a_n Q_0(y).$$

Example 5.4.1. *Find the zeros of polynomials of the form*

$$P(x) = ax^4 + bx^3 + cx^2 + bx + a.$$

Solution: The key is to 'fold' the reciprocal polynomials so we can combine the two a coefficients and the two b coefficients. So we divide $P(x)$ by x^2 to get

$$a(x^2 + x^{-2}) + b(x + x^{-1}) + c.$$

We will always divide a reciprocal polynomial of degree $2n$ by x^n just so we can obtain a polynomial similar to the previous results.

Now we can apply Lemma 5.4.1 to write this equation as

$$a(y^2 - 2) + by + c = ay^2 + by + (c - 2a),$$

where $y = x + x^{-1}$. This quadratic in y has the zeros

$$y_1, y_2 = \frac{-b \pm \sqrt{b^2 - 4a(c - 2a)}}{2a}.$$

Now $y = x + x^{-1}$, so the zeros of $P(x)$ are the zeros of

$$x^2 - y_1 x + 1 = 0, \quad x^2 - y_2 x + 1 = 0$$

that is,

$$\frac{y_1 \pm \sqrt{y_1^2 - 4}}{2}, \quad \frac{y_2 \pm \sqrt{y_2^2 - 4}}{2}.$$

5. Elementary Techniques for Finding Zeros

For example, if $a = b = c = 1$, then $y_1, y_2 = (-1 \pm \sqrt{5})/2$; so the zeros of $x^4 + x^3 + x^2 + x + 1$ are

$$\frac{-1+\sqrt{5}}{4} \pm \frac{\sqrt{10+2\sqrt{5}}}{4}i, \quad \frac{-1-\sqrt{5}}{4} \pm \frac{\sqrt{10-2\sqrt{5}}}{4}i$$

∎

Lemma 5.4.3. *If $P(x)$ is a reciprocal polynomial with odd degree, then $P(-1) = 0$. Further, $Q(x) = P(x)/(x+1)$ is a reciprocal polynomial of degree $n-1$.*

Proof. $P(x)$ is reciprocal so $P(x) = x^n P(1/x)$, in particular

$$P(-1) = (-1)^n P(-1) = -P(-1),$$

thus $2P(-1) = 0$. To see that $Q(x)$ is reciprocal, we note:

$$x^{n-1}Q(1/x) = \frac{x^n}{x} \cdot \frac{P(1/x)}{\frac{1}{x}+1} = \frac{P(x)}{x+1} = Q(x).$$

∎

Example 5.4.2. *Find the zeros of*

$$P(x) = 2x^5 + 3x^4 - 5x^3 - 5x^2 + 3x + 2.$$

Solution: We know -1 is a zero, so we remove this zero by synthetic division as follows:

$$\begin{array}{r|rrrrrr} -1 & 2 & 3 & -5 & -5 & 3 & 2 \\ & & -2 & -1 & 6 & -1 & -2 \\ \hline & 2 & 1 & -6 & 1 & 2 & 0 \end{array}$$

So the remaining zeros of $P(x)$ are the zeros of

$$Q(x) = 2x^4 + x^3 - 6x^2 + x + 2.$$

This has degree four, so dividing by x^2 and setting $y = x + x^{-1}$, this becomes

$$2(y^2 - 2) + 1(y) - 6(1) = 2y^2 + y - 10 = (2y+5)(y-2).$$

Thus the zeros of $Q(x)$ are the zeros of $x^2 - 2x + 1$ and $x^2 + (5/2)x + 1$ which are 1, 1, -2, and $-1/2$. So the zeros of $P(x)$ are

$$-1, \quad 1, \quad 1, \quad -2, \quad -\frac{1}{2}.$$

∎

5.4. Reciprocal Polynomials

Exercises 5.4:

1. Prove Lemma 5.4.1 by showing $Q_{m+1}(y) = yQ_m(y) - Q_{m-1}(y)$.

2. Prove the first and third conditions given in the definition of reciprocal polynomial are equivalent.

3. Prove the first and second conditions given in the definition of reciprocal polynomial are equivalent.

4. Find the zeros of the following polynomials.

    a) $x^3 + x^2 + x + 1$

    b) $x^3 + 2x^2 + 2x + 1$

    c) $x^3 + bx^2 + bx + 1$

    d) $x^6 + bx^3 + bx^2 + 1$

    e) $2x^3 + 3x^2 + 3x + 2$

    f) $ax^3 + bx^2 + bx + a$

    g) $ax^{3n} + bx^{2n} + bx^n + a$

    h) $x^4 + x^3 + x + 1$

    i) $x^4 + 4x^3 + 6x^2 + 4x + 1$

    j) $x^4 + 2x^3 + 3x^2 + 2x + 1$

    k) $2x^4 - x^3 + x^2 - x + 2$

    l) $2x^4 + 5x^3 + 6x^2 + 5x + 2$

    m) $ax^{4n} + bx^{3n} + cx^{2n} + bx^n + a$

    n) $x^5 + x^4 + x^3 + x^2 + x + 1$

    o) $x^5 + 2x^4 + 3x^3 + 3x^2 + 2x + 1$

    p) $2x^5 + x^4 + x + 2$

    q) $2x^5 - 3x^4 + 2x^3 - 2x^2 + 3x - 2$

    r) $ax^5 + bx^4 + cx^3 + cx^2 + bx + a$

    Hint: For (q) divide by $x - 1$.

5. In this problem we will find the zeros of $x^7 - 1$.

    a) Clearly one is a zero, so divide it out to find a sixth-degree polynomial $P(x)$.

    b) Use the technique of this section to find the associated cubic $Q(y)$.

    c) Solve $Q(y)$ using Cardan's Formula or Maple.

    d) Now use your solution to $Q(y) = 0$ to find the six zeros of $P(x)$ in terms of radicals (not $\cos \frac{\pi}{7} \ldots$). Maple may help.

5. Elementary Techniques for Finding Zeros

6. (*) A polynomial $P(x)$ of degree $n > 1$ is called a **reciprocal polynomial of the second class** if $P(x) = -x^n P(1/x)$. Show that the following two conditions are each equivalent to this definition.

   a) $a_0 = -a_n$ and if r is a zero of P, then $1/r$ is also a zero.

   b) $a_0 = -a_n$, $a_1 = -a_{n-1}$, $a_3 = -a_{n-3}$, $\ldots$

   A reciprocal polynomial is also called a **reciprocal polynomial of the first class**.

7. (*) Prove the following.

   > **Theorem 5.4.4.** *Let $P(x)$ be a reciprocal polynomial of the second class. $P(x)$ has the zero $x = 1$, and $P(x)/(x-1)$ is a reciprocal polynomial of the first class.*

8. Show a reciprocal polynomial of the second class with degree $2n+2$ or $2n+1$ may be solved by solving a polynomial of degree n, and then solving n quadratic polynomials.

9. We might attempt to generalize the concept of reciprocal polynomials by calling a polynomial c-reciprocal if it satisfies the condition: $a_0 = c \cdot a_n$ and if r is a zero of P, then $1/r$ is also a zero.

   a) Show an equivalent condition is $P(x) = cx^n P(1/x)$.

   b) Show a reciprocal polynomial of the first class is 1−reciprocal (see problem 6), and that a reciprocal polynomial of the second class is (-1)−reciprocal.

10. (*) Show that there are no c−reciprocal polynomials if $c \neq \pm 1$. Hint: Let $P(x)$ be a c−reciprocal polynomial, use part (a) of the previous problem to show the following sequence of results.

    a) If $c \neq 1$, then $P(1) = 0$

    b) If $c \neq \pm 1$, then $P(-1) = 0$

    c) If $c \neq \pm 1$, then $x^2 - 1$ divides $P(x)$

    d) If $c \neq \pm 1$, then $P(x)/(x^2 - 1)^2$ is c−reciprocal

    e) If $c \neq \pm 1$, then $P(x)/(x^2 - 1)$ is $(-c)$−reciprocal

    f) If $c \neq \pm 1$, then $(x^2 - 1)^2$ divides $P(x)$

    This leads to a contradiction because _____.

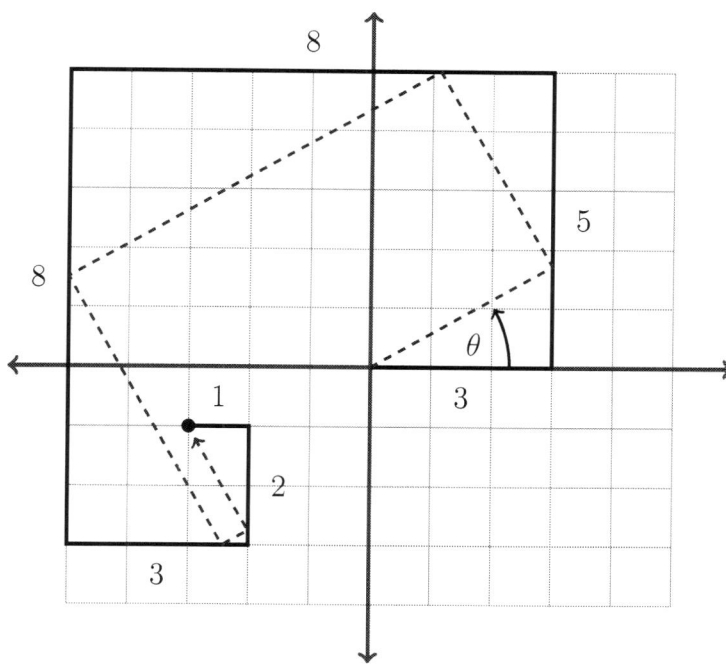

Figure 5.3: Lill's Method for $3x^6 + 5x^5 + 8x^4 + 8x^3 + 3x^2 + 2x + 1$

5.5 Lill's Method for Solving Polynomials

Lill's method of solving polynomials is an amusing mixture of geometry and billiards. It is not a practical approach for most polynomials, but it is intriguing in many ways. It was first described by the Austrian engineer Eduard Lill in 1867 in [5] and [6].

The first step in Lill's method is to draw a "billiard table" by repeatedly:

> moving forward some distance and then turning left 90°.

For example, to solve

$$3x^6 + 5x^5 + 8x^4 + 8x^3 + 3x^2 + 2x + 1 = 0,$$

we begin at the origin, move 3 units along the x-axis (because the first coefficient is 3 and then turn 90° left (please follow along in Figure 5.3). Next go 5 units forward (the next coefficient) and turn, 8 units then turn, 8 units then turn, 3 units then turn, 2 units then turn, and finally go 1 unit to find the terminal point. The next step is to "shoot a billiard ball" from the origin at some angle θ with the x-axis. The ball must be shot so that it will bounce off each of these "walls" we have drawn in turn and finally end up at our terminal point. Lill showed that $-\tan\theta$ is a zero of our polynomial!

5. Elementary Techniques for Finding Zeros

Lill's method will find all of the real zeros of a polynomial, so there are as many choices of θ as there are distinct real zeros. The eighth degree polynomial above has two real zeros, and the path for the other is shown in Figure 5.9.

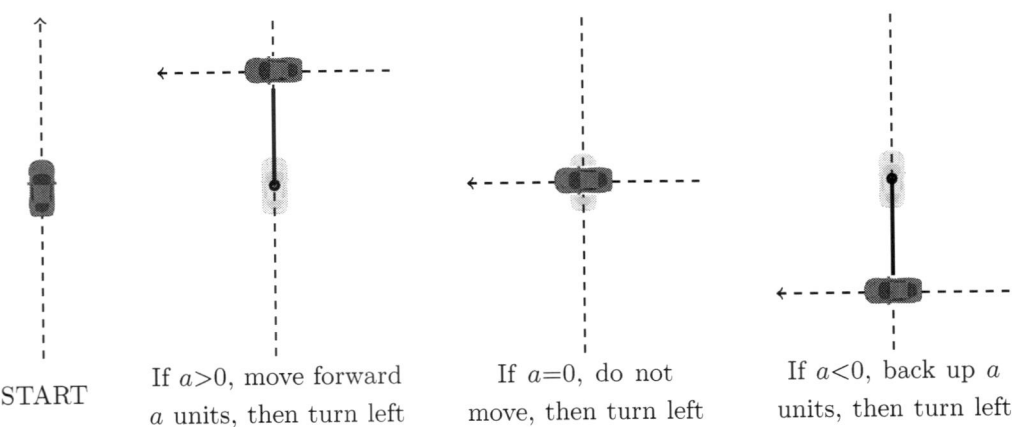

Figure 5.4: When Drawing the Side for the Coefficient a: Always Turn Left

We oversimplified things a bit above, so let's be a bit more careful here.

- **Drawing Walls:** When drawing the table, we always turn left; and once we are facing left the distance we travel is decided by the coefficient. In particular, when a coefficient is negative, we do not turn around, we just move $-a$ units forward by backing up a units. We still are facing the same direction, we just move in reverse (see Figure 5.4). If the coefficient is zero, we do not move at all (see also exercise 4.)

- **Solving:** When we are seeking a solution, we view the segments of the "billiard table" as extended into lines, and then when we turn 90° **we always turn in the direction necessary to intersect the next segment**. (So always left when drawing, but we turn which ever direction works when solving.) See the left image in Figure 5.5 for an example of both a negative coefficient and extended walls. Even if the wall has length zero, we can extend it into a line the direction we are facing (see Figure 5.8).

You might want to scan several of the figures to make sure you understand these points.

Instead of a formal general proof, we will show that Lill's method works using the cubic $ax^3 + bx^2 - cx + d$ (with a, b, c and d positive) which is illustrated in Figure 5.5. The proof is based on the fact that each of the triangles is similar, so each has the angle θ in it.

First, suppose the path meets the wall for b (the line $x = a$) at the point (a, q) (see the righthand image in Figure 5.5). So using the triangle with a vertex at the

5.5. Lill's Method for Solving Polynomials

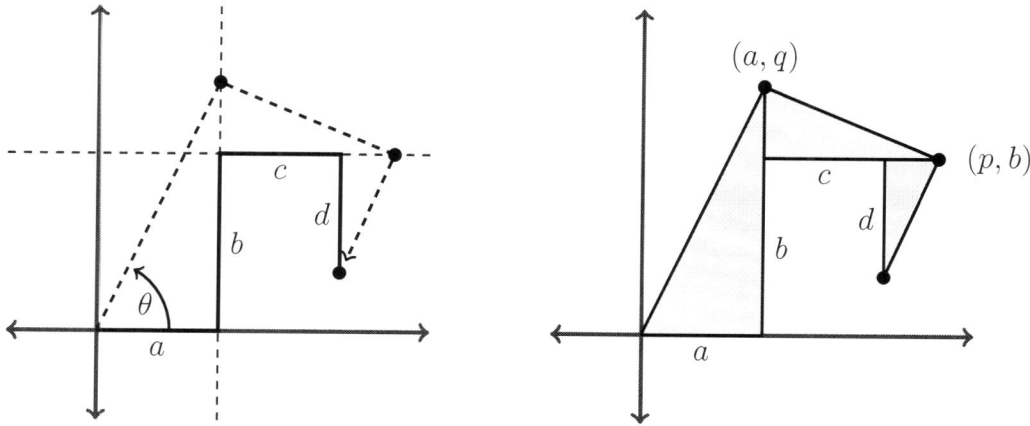

Figure 5.5: Lill's Method for the Cubic $ax^3 + bx^2 - cx + d$ with $a, b, c > 0$

origin we have
$$\tan\theta = \frac{q}{a},$$
so $a\tan\theta = q$. Next, from the triangle on top we have
$$\tan\theta = \frac{p-a}{q-b} = \frac{p-a}{a\tan\theta - b}$$
so $p - a = a\tan^2\theta - b\tan\theta$. From the third triangle we have
$$\tan\theta = \frac{d}{p-a-c} = \frac{d}{a\tan^2\theta - b\tan\theta - c},$$
so $a\tan^3\theta - b\tan^2\theta - c\tan\theta = d$, or
$$a(-\tan\theta)^3 + b(-\tan\theta)^2 - c(-\tan\theta) + d = 0.$$

This shows that $x = -\tan\theta$ is a zero of $ax^3 + bx^2 - cx + d$. It also shows us that Lill's method is closely related to Horner's Process (Theorem 6.2.2).

It is a very pleasant exercise to implement Lill's method using GeoGebra—give it a try! Note that Lill's method for the general cubic can be also done using origami (just paper and folds), see section 8.6.

How do we Know What Angle to Use?

Usually we do not know what angle to use—finding the angle is equivalent to finding a zero of the polynomial. If we do this in a program like GeoGebra, we can control the angle with a slider, and just slide along until we get an angle that will work. There are however the two easiest cases: linear and quadratic. These we can do geometrically.

5. Elementary Techniques for Finding Zeros

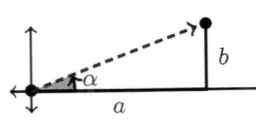

For the linear polynomial $ax + b$, the angle in Lill's method is obvious—just aim for the final point. The negative of the tangent of this angle is $-\frac{b}{a}$, which is clearly the zero of the polynomial.

For quadratics, there is also an easy geometric way to see where the paths should go. Draw the circle which has the line segment from the starting to the finishing point as a diameter (see the figure on the right and Figure 5.7 for different views). This circle intersects the line defining the middle of the three "walls" where the solution paths bounce off this wall. It is clear geometrically that these paths bounce of this line (and the circle) in a way which forms right angles and then end at the finishing point. To show that $-\tan\alpha$ is a zero of the quadratic in each case takes a little algebra–we'll do so below.

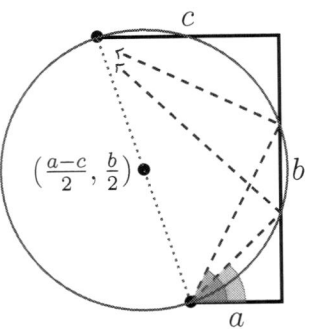

Figure 5.6.

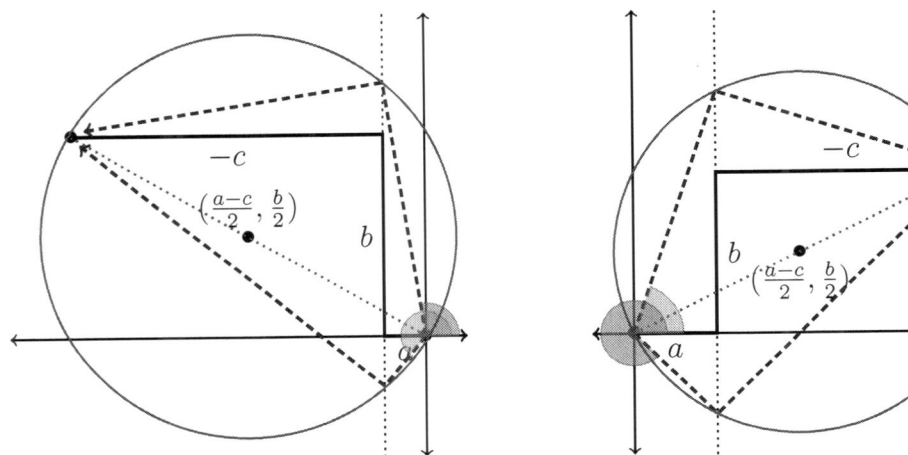

Figure 5.7: Lill's Method for $-ax^2 + bx + c$ and $ax^2 + bx - c$ with $a, b, c > 0$

Since the starting point was the origin $(0,0)$, and the second "wall" is on the line $x = a$; call one of the intersection points (a, y). Then the square of the radius of the circle is given by both sides of the following equation.

$$\left(a - \frac{a-c}{2}\right)^2 + \left(y - \frac{b}{2}\right)^2 = \left(0 - \frac{a-c}{2}\right)^2 + \left(0 - \frac{b}{2}\right)^2$$

This simplifies to
$$y = \frac{b \pm \sqrt{b^2 - 4ac}}{2}.$$

5.5. Lill's Method for Solving Polynomials

So the negative of the tangent of the starting angle is

$$\frac{-b \pm \sqrt{b^2 - 4ac}}{2a};$$

which we recognize as the quadratic formula! It is possible that this circle is to small to intersect that wall-line; but this happens when $b^2 - 4ac < 0$ and the quadratic has no solution to find.

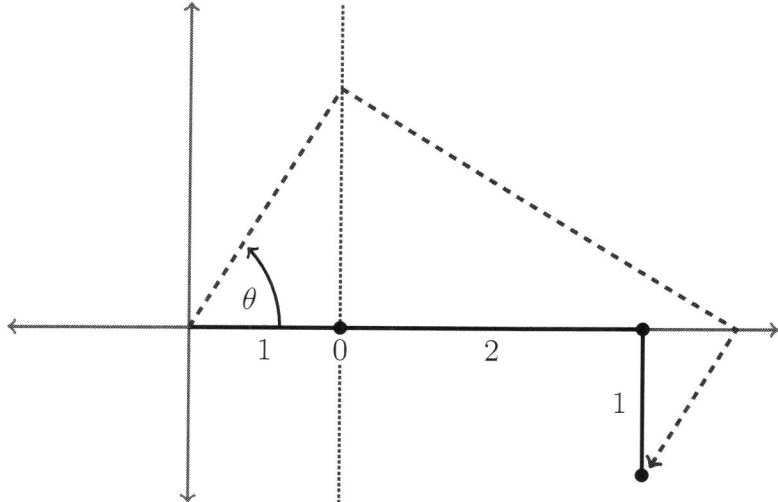

Figure 5.8: Lill's Method for $x^3 - 2x + 1$

Exercises 5.5:

1. Use GeoGebra to solve the the quadratic $ax^2 + bx + c = 0$. Hint: To start create three sliders to define the coefficients a, b and c. Enter the points (0,0), (a,0), (a,b) and (a-c,b) one at a time in the input bar (these will now move with the sliders). Draw the three lines on which the sides of length a, b and c will lie. Place another point on the plane that you'll use to define the ray making the angle θ. ...

2. Use GeoGebra to solve the general cubic $ax^3 + bx^2 + cx + d = 0$.

3. What happens to Lill's method when it has no real solution?

4. When a coefficient is zero, we move 0 units, mark the direction with a dashed line, and turn again as usual. In Figure 5.8 we draw this for $x^3 - 2x + 1$.

    a) Which of the three solution is found in Figure 5.8?

    b) Draw the two paths that come from the other two solutions.

5. Elementary Techniques for Finding Zeros

c) For $ax^3 - bx + c$, use a drawing like Figure 5.8 to prove Lill's method finds a zero of the cubic.

5. By hand apply Lill's method to solve $x^3 - x^2 - 3x - 3$. Plot the paths for each of the three solutions. Hint: Carefully draw the paths defined by the polynomial on a large grid. The zeros have been chosen so that the triangles are easy to draw accurately. For the solution $x = 1$, think carefully about what happens at the point $(1, -1)$.

6. By hand apply Lill's method to solve $6x^3 - 3x^2 - 2x + 1$. Plot the paths for each of the three solutions. Hint: The zeros have been chosen so that the triangles are easy to draw accurately.

7. One solution to $P(x) = 3x^6 + 5x^5 + 8x^4 + 8x^3 + 3x^2 + 2x + 1$ is illustrated in Figure 5.3. Another is illustrated in Figure 5.9. Call this solution r.

    a) What is r? Hint: this is easy!

    b) Look at the illustration on the right of Figure 5.9. If we rotated this path so the first segment was parallel to the x-axis, what polynomial would it represent?

    c) Compare your answer to the previous part to $P(x)/(x - r)$.

    d) Explain how the dashed-path in Figure 5.9 can be used to find the second solution of $P(x)$.

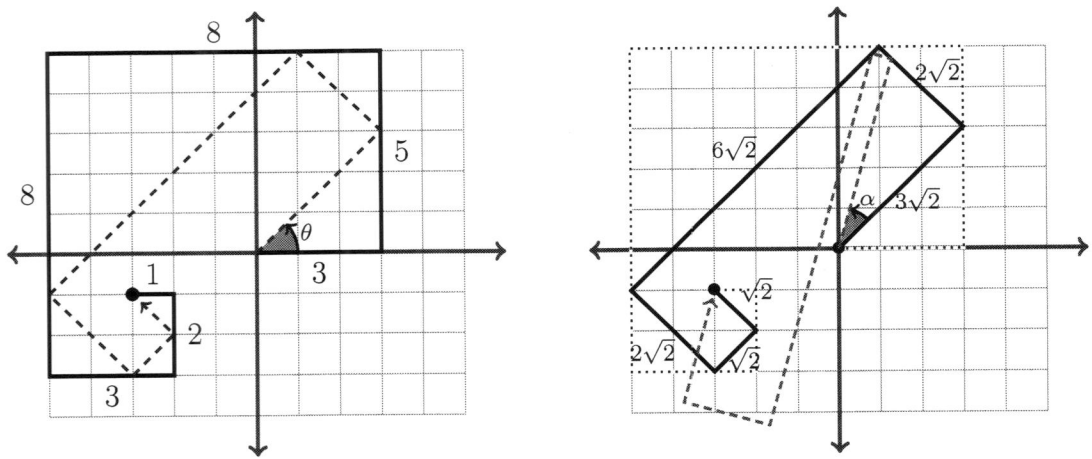

Figure 5.9: Lill's Method for $3x^6 + 5x^5 + 8x^4 + 8x^3 + 3x^2 + 2x + 1$

8. Suppose that $-\tan\theta$ corresponds to one solution of $ax^2 + bx + c$ as denoted by the path in Figure 5.10. Show that $-\tan\alpha$ (where α is defined by the path and the line through the starting and ending point) is the other solution. Hint: Perhaps similar triangles (see Figure 5.5). A brute force method would be to suppose $-\tan\beta$ is the other zero, find these tangents from the quadratic formula, and then use the angle addition formula for tangents to check if $\tan(\alpha + \theta) = b/(a - c)$.

5.5. Lill's Method for Solving Polynomials

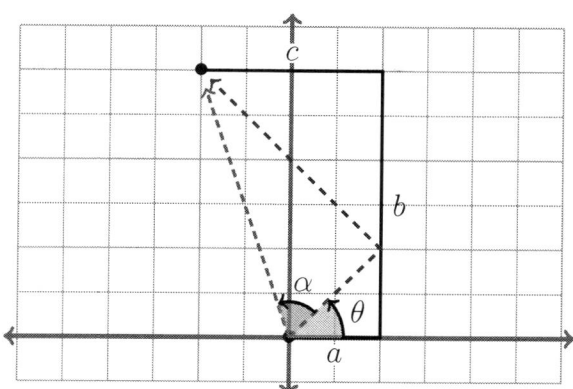

Figure 5.10: Lill's Method for $ax^2 + bx + c$ (Problem 8)

Relationships Between the Zeros and the Coefficients

Number rules the universe.
The Pythagoreans (c. 500 BC)

God ever geometrizes.
Plato (c. 380 BC)

God ever arithmetizes.
C. G. J. Jacobi (c. 1830)

As far as the mathematical theorems refer to reality, they are not sure, and as far as they are sure, they do not refer to reality.
Albert Einstein

Chapter 6

Relationships Between the Zeros and the Coefficients

In this chapter we study the important relationship between the zeros and coefficients of polynomials. We use these relationships to help find zeros, to gain information about the zeros, and even to find missing coefficients of polynomials. We will study how to manipulate the zeros of polynomials, forming new polynomials whose zeros are related in some manner to the old.

Remember the discriminant of the quadratic $b^2 - 4ac$? This is zero if and only if the quadratic $ax^2 + bx + c$ has a double zero. In this chapter we will study symmetric functions with which we could find such a discriminant for polynomials of any degree. Finally, we will recall Newton's Identities, which can be used to find the sums of powers of zeros of any polynomial.

6.1 The Coefficients

Let $r_1, r_2, \ldots, r_n$ be the zeros of the polynomial

$$P(x) = a_0 x^n + a_1 x^{n-1} + \ldots + a_n.$$

By the first corollary to the Fundamental Theorem of Algebra we have the identity

$$P(x) = a_0(x - r_1)(x - r_2) \ldots (x - r_n).$$

Multiply out the right hand side of this identity to find:

$$a_0 x^n - a_0(r_1 + r_2 + \ldots + r_3)x^{n-1} + \ldots + (-1)^n a_0 r_1 r_2 \ldots r_n.$$

6. Relationships Between the Zeros and the Coefficients

This polynomial is identical to the original polynomial $P(x)$, so their coefficients are equal. Equating coefficients and dividing by a_0 we find:

$$-\frac{a_1}{a_0} = r_1 + r_2 + \ldots + r_n$$

$$\frac{a_2}{a_0} = r_1 r_2 + r_1 r_3 + \ldots + r_{n-1} r_n$$

$$-\frac{a_3}{a_0} = r_1 r_2 r_3 + r_1 r_2 r_4 + \ldots + r_{n-2} r_{n-1} r_n$$

$$\vdots \qquad \vdots$$

$$(-1)^n \frac{a_n}{a_0} = r_1 r_2 \ldots r_n$$

To translate these equations into English we say that a_1/a_0 is the sum of the zeros of P, $-a_2/a_0$ is the sum of the products of the zeros taken two at a time, a_3/a_0 is the sum of the products of the zeros taken three at a time....

Theorem 6.1.1. *If $r_1, r_2, \ldots, r_n$ are the zeros of the polynomial*

$$P(x) = a_0 x^n + a_1 x^{n-1} + \ldots + a_n$$

with $a_0 \neq 0$, then $(-1)^m a_m/a_0$ is the sum of the products of the zeros taken m at a time.

Example 6.1.1. *If $P(x)$ has the simple zeros $2, -3, 7$, and is monic, find $P(x)$.*

Solution: $P(x)$ has three zeros, so $P(x)$ is a cubic, say

$$P(x) = a_0 x^3 + a_1 x^2 + a_2 x + a_3.$$

We know $P(x)$ is monic, so $a_0 = 1$. The other coefficients satisfy

$$\begin{aligned}
-a_1 &= (2) + (-3) + (7) &&= 6, \\
a_2 &= (2)(-3) + (2)(7) + (-3)(7) &&= -13, \\
-a_3 &= (2)(-3)(7) &&= -42.
\end{aligned}$$

Hence $P(x) = x^3 - 6x^2 - 13x + 42$. ∎

Example 6.1.2. *For each of the following polynomials find (a) the sum of the zeros; (b) the product of the zeros; and (c) the sum of the products taken two at a time.*

$$x^3 - 27x + 18$$
$$5x^4 + 6x^3 + 3x^2 - 7x + 23$$
$$-8x^7 + 234x^6 - 25x^3 + 47x^2 - 123x$$
$$x^5 - 4x^4 + 234x^3 + 496x^2 - 782x + 3956$$

6.1. The Coefficients

Solution: (a) The sum of the zeros is $-a_1/a_0$, so for these polynomials the sum is respectively: 0, $-6/5$, $234/8 = 117/4$, and 4. (b) The product of the zeros is $(-1)^n a_n/a_0$ or -18, $23/5$, 0 and -3956 respectively. Finally, (c) the sum of the products of the zeros taken two at a time is a_2/a_0 or -27, $3/5$, 0, and 234. ∎

Example 6.1.3. *Find the sum of the squares of the zeros of*

$$3x^4 - 2x^3 + 25x^2 - 161x + 2.$$

Solution: Let the zeros of this polynomial be r_1, r_2, r_3, r_4. The sum of the zeros, and the sum of the products of the zeros taken two at a time are:

$$r_1 + r_2 + r_3 + r_4 = \frac{2}{3}; \qquad r_1r_2 + r_1r_3 + r_1r_4 + r_2r_3 + r_2r_4 + r_3r_4 = \frac{25}{3}.$$

Notice:

$$(r_1 + r_2 + r_3 + r_4)^2 = r_1^2 + r_2^2 + r_3^2 + r_4^2 + 2(r_1r_2 + r_1r_3 + r_1r_4 + r_2r_3 + r_2r_4 + r_3r_4)$$

which using the information above is

$$\left(\frac{2}{3}\right)^2 = r_1^2 + r_2^2 + r_3^2 + r_4^2 + 2\left(\frac{25}{3}\right).$$

The sum of the squares of the zeros is then $\left(\frac{2}{3}\right)^2 - 2\left(\frac{25}{3}\right) = -\frac{146}{9}$. (See also exercise 3.) Notice that the sum of the squares is negative, so the polynomial must have complex zeros. ∎

Example 6.1.4. *Find the zeros of the following polynomial given that it has two double zeros.*

$$x^4 - 2x^3 - 3x^2 + 4x + 4$$

Solution: Let the zeros be r, r, s, s. Then the sum and the product of these zeros are 2 and 4 respectively, i.e., $2r + 2s = 2$, and $r^2 s^2 = 4$. From the second, we see $rs = \pm 2$. Using this to eliminate r in the first, we see $s^2 - s \pm 2 = 0$; so the zeros r, r, s, s must be $-1, -1, 2, 2$. ∎

Example 6.1.5. *Let $P(x) = x^3 + 7x^2 - x + N$ have the zeros r_1, r_2, r_3. Given that $r_1 = -r_2$, find the zeros and the missing coefficient N.*

Solution: We know that $r_1 + r_2 = 0$, and that $r_1 + r_2 + r_3 = -7$, so $r_3 = -7$. Next $r_1r_2 + r_1r_3 + r_2r_3 = -1$, so $r_1r_2 = -r_1^2 = -1$, and $r_1r_2r_3 = -N$, so $N = -7$. Thus the zeros are $1, -1, -7$ and the coefficient N is -7. (Once we knew one zero was -7, we could have used synthetic division and the Factor Theorem to find N.) ∎

6. Relationships Between the Zeros and the Coefficients

Exercises 6.1:

1. For each of the following polynomials find the sum of the zeros; the sum of the products of the zeros taken two, and then three at a time; and finally the products of the zeros.

   a) $x^3 - 4x^2 + 3x + 12$

   b) $2x^4 - 15x^2 + 6x + 23$

   c) $3x^3 + 3x^2 + 9$

   d) $-6x^5 + 16x^2 + 27x - 18$

   e) $-1000x^3 + 27000x^2 - 18000x + 345000$

   f) $x^{1000} + 14x^2 - 45x - 1000000$

   g) $x^{9999} + 14x^2 - 45x - 1000000$

   h) $x^2 - 1$

   i) $x^n - 1 \quad (n > 2)$

   j) $3x^n - 2x^{n-1} + 3x^{n-2} - 1 \quad (n > 2)$

2. Find the zeros of the following polynomials and determine the missing coefficient(s) using the given information about the zeros. (Note: N and M are real numbers.)

   a) $x^3 - 5x^2 - 4x + N, \quad r_1 = -r_2$

   b) $4x^3 - 13x^2 - 8x + N, \quad r_1 + r_2 = 4$

   c) $x^3 - x^2 + Nx + 39, \quad r_1 + r_2 = 4$

   d) $2x^3 + x^2 - 13x + N, \quad r_1 = 4r_2$

   e) $x^4 - 14x^3 + 69x^2 + Nx + M, \quad r_1 = r_2, r_3 = r_4$

   f) (*) $8x^4 - 28x^3 + 18x^2 + Nx - 27, \quad r_1 = r_2 = r_3$
   Hint: Use the coefficients of x^3 and x^2 to get two possible values for r_1; this gives values for r_4. Use the constant term to find the correct values.

   g) $x^3 + x^2 + Nx + M, \quad r_1 = 1 + i$

   h) $x^3 + Nx^2 + Mx + 4, \quad r_1 = 1 + i$

   i) $10x^3 + Nx + M, \quad r_1 = 4 - 5i$

   j) $x^3 - 7x^2 + Nx - 63, \quad r_1$ pure imaginary

   k) $2x^3 + Nx^2 + 2x + M, \quad r_1$ pure imaginary

3. Let $P(x) = a_0 x^n + a_1 x^{n-1} + a_2 x^{n-2} + \ldots$ Show that the sum of the squares of the zeros of P is $(a_1^2 - 2a_0 a_2)/a_0^2$.

4. For each polynomial in problem one, find the sum of the squares of the zeros. Hint: Use the previous problem.

6.1. The Coefficients

5. (*) Let $P(x) = a_0 x^n + a_1 x^{n-1} + a_2 x^{n-2} + a_3 x^{n-3} + \ldots$ Show that the sum of the cubes of the zeros of P is $(3a_0 a_1 a_2 - 3a_0^2 a_3 - a_1^3)/a_0^3$.

6. For each polynomial in problem 1, find the sum of the cubes of the zeros. Hint: Use the previous problem.

7. Let $P(x) = a_0 x^n + a_1 x^{n-1} + \ldots + a_n$. Prove that $-a_{n-1}/a_n$ is the sum of the reciprocals of the zeros of P.

8. For each polynomial in problem 1, find the sum of the reciprocals of the zeros.

9. Let $P(x) = 3x^{97} - 4x^{96} + 25x^{95} + \ldots$ Prove that $P(x)$ has at least two imaginary zeros. Hint: What is the sum of the squares of the zeros?

10. Let $n > 1$ be an integer. Show that the sum of the nth zeros of any number is equal to zero. Hint: What equation do they satisfy?

11. Let x_1, x_2, x_3, x_4 be the x-values of the points where the graph of the line $ax + b$ intersects the graph of $3x^4 - 15x^3 + 17x^2 + 23x + 7$. Find $x_1 + x_2 + x_3 + x_4$.

12. The equation $x^3 - 2x^2 + \ldots = 0$ has real coefficients and the complex zero $2 + 3i$. Find the other coefficients.

13. Show that if r is a zero of $x^3 + ax^2 + bx + c$, then the two remaining zeros satisfy $x^2 + (a+r)x - \frac{c}{r} = 0$.

14. Suppose the polynomial $x^4 + ax^3 + bx^2 + cx + d$ has the zeros r and s. Find the quadratic equation the two remaining zeros satisfy (do not use b or c in your answer).

15. Let a, b, c be real numbers for which $a+b+c = 3$, $a^2+b^2+c^2 = 5$, and $a^3+b^3+c^3 = 7$. Find $a^4 + b^4 + c^4$. Hint: Let $P(x) = (x-a)(x-b)(x-c)$.

16. The zeros of the following polynomials are in arithmetic progression. Find the zeros and the value of the unknown coefficient.

    a) $x^3 - 6x^2 + Nx + 10$

    b) $x^3 - 12x^2 + Mx - 28$

    Hint: Call the zeros $r-d, r$ and $r+d$. The problem is now easy if you use what you know about the second and last coefficients.

17. Prove that if the zeros of $x^3 - ax^2 + bx - c$ form an arithmetic sequence, then $2a^3 - 9ab + 27c = 0$. Hint: Call the zeros $r - d, r$ and $r + d$; evaluate $2a^3 - 9ab + 27c$ in terms of r and d.

18. (*) Find the zeros of the following polynomials given that the zeros are in arithmetic progression.

    a) $x^5 - 15x^4 - 35x^3 + $???

b) $x^7 - 7x^6 - 483x^5 + ???$

Sadly a dog ate the rest of the polynomials, but you can still do this problem!

19. You woke up late in algebra class just in time to hear the teacher say "the zeros of this polynomial are all prime," but you only had time to write down the constant term 30030. Determine the polynomial and all of its zeros.

6.2 Transforming the Zeros of Polynomials I

When finding the zeros of a polynomial $P(x)$ it is often useful to consider a second polynomial $Q(x)$ whose zeros bear some relation to those of the first. We say that $Q(x)$ is obtained by **transforming** $P(x)$. We begin with a few of the easier transformations.

> **Theorem 6.2.1 (Multiplying the Zeros by a Constant).** Let $r_1, r_2, \ldots, r_n$ be the zeros of the polynomial
>
> $$P(x) = a_0 x^n + a_1 x^{n-1} + \ldots + a_n \quad \text{with} \quad a_0 \neq 0.$$
>
> Then $r_1 m, r_2 m, \ldots, r_n m$, are the zeros of
>
> $$Q(x) = a_0 x^n + a_1 m^1 x^{n-1} + a_2 m^2 x^{n-2} + a_3 m^3 x^{n-3} + \ldots + a_n m^n.$$

Example 6.2.1. Let $P(x) = 3x^4 + 2x^2 - 6x + 7$. Find the polynomials whose zeros are: five times the zeros of $P(x)$; the negative of the zeros of $P(x)$; and -2 times the zeros of $P(x)$.

Solution: We use the rule for multiplying the zeros by a constant, using $m = 5, -1$ and -2 respectively to find the desired polynomials (notice that the coefficient of x^3 is zero).

$$\begin{aligned} m = 5: & \quad 3x^4 + 50x^2 - 750x + 4375 \\ m = -1: & \quad 3x^4 + 2x^2 + 6x + 7 \\ m = -2: & \quad 3x^4 + 8x^2 + 48x + 112. \end{aligned}$$

∎

Another important transformation is **translation**, that is, forming a polynomial $Q(x)$ whose zeros are the zeros of $P(x)$ minus the constant c. It is easy to see that such a polynomial is given by

$$Q(x) = P(x + c) = a_0(x + c)^n + a_1(x + c)^{n-1} + a_2(x + c)^{n-2} + \ldots + a_n.$$

6.2. Transforming the Zeros of Polynomials I

It is possible to write $Q(x)$ in the standard form by multiplying each of these terms out and grouping like powers, but there is a much simpler method.

Theorem 6.2.2 (Horner's Process). *Let $P(x)$ be a polynomial with degree n, then*
$$P(x+c) = b_0 x^n + b_1 x^{n-1} + \ldots + b_n.$$
where b_n are defined recursively as follows.
$$\begin{aligned} P(x) &= (x-c)Q_1(x) + b_n \\ Q_1(x) &= (x-c)Q_2(x) + b_{n-1} \\ &\vdots \\ Q_n(x) &= (x-c)Q_{n+1}(x) + b_0 \end{aligned}$$
That is, the b_n are the remainders upon successive divisions of $P(x)$ by $(x-c)$.

This theorem may look complicated, but the application as simple as repeatedly applying synthetic division. Let's show this with an example.

Example 6.2.2. *Let $P(x) = 5x^4 - 14x^3 + 3x^2 - 6x + 7$. Find a polynomial $Q(x)$ such that the zeros of $Q(x)$ are 3 less than the zeros of $P(x)$.*

Solution: Using synthetic division, the required sequence of divisions are easily performed as in Figure 6.1. So we see $Q(x) = 5x^4 + 46x^3 + 147x^2 + 174x + 43$. ∎

```
3 | 5   -14    3    -6    7
         15    3    18    36
    5    1     6    12  | 43    ⇐ b_4
         15    48   162
    5    16    54 | 174          ⇐ b_3
         15    93
    5    31  | 147                ⇐ b_2
         15
    5  | 46                       ⇐ b_1
    ⇑
    b_0
```

Figure 6.1: Horner's Process, Example 6.2.2

A third type of transformation is to form a polynomial $Q(x)$ whose zeros are the reciprocals of $P(x)$.

101

6. Relationships Between the Zeros and the Coefficients

Theorem 6.2.3 (Inverting the Zeros). *Let $P(x)$, $Q(x)$ be defined as follows*

$$P(x) = a_0 x^n + a_1 x^{n-1} + \ldots + a_{n-1} x + a_n$$

$$Q(x) = x^n P\left(\frac{1}{x}\right) = a_n x^n + a_{n-1} x^{n-1} + \ldots + a_1 x + a_0$$

with $a_0 a_n \neq 0$. Then the zeros of $Q(x)$ are the reciprocals of the zeros of $P(x)$.

Notice that all the theorem asks us to do is reverse the order of the coefficients.

Before we give an application of this theorem, let's look at a simple illustration. Suppose that the zeros of the monic polynomial $P(x)$ are r_1, r_2 and r_3. Then

$$P(x) = (x - r_1)(x - r_2)(x - r_3)$$
$$= x^3 - (r_1 + r_2 + r_3)x^2 + (r_1 r_2 + r_1 r_3 + r_2 r_3)x - r_1 r_2 r_3,$$

so

$$x^3 P\left(\frac{1}{x}\right) = x^3 \left(\frac{1}{x} - r_1\right)\left(\frac{1}{x} - r_2\right)\left(\frac{1}{x} - r_3\right)$$
$$= (1 - r_1 x)(1 - r_2 x)(1 - r_3 x)$$

$$x^3 P\left(\frac{1}{x}\right) = -r_1 r_2 r_3 x^3 + (r_1 r_2 + r_1 r_3 + r_2 r_3)x^2 - (r_1 + r_2 + r_3)x + 1.$$

So as the theorem stated, the coefficients were just reversed.

Example 6.2.3. *Let $P(x) = 3x^3 + 7x + 2$ have the zeros r_1, r_2, r_3. Find a polynomial with the zeros*

$$1 + \frac{1}{r_1}, \quad 1 + \frac{1}{r_2}, \quad 1 + \frac{1}{r_3}.$$

Solution: We first find a polynomial whose zeros are the reciprocals of the zeros of P, and then translate by one. The polynomial whose zeros are the reciprocals of the zeros of P is

$$x^3 P(x^{-1}) = 2x^3 + 7x^2 + 3.$$

We now add one to each zero using Horner's process (shown in Figure 6.2). We can see the desired polynomial is

$$2x^3 + x^2 - 8x + 8.$$

∎

6.2. Transforming the Zeros of Polynomials I

```
-1 | 2   7    0   3
   |    -2   -5  5
   ------------------
     2   5   -5 | 8
         -2   -3
     ---------------
     2   3  | -8
         -2
     --------
     2  | 1
```

Figure 6.2: Horner's Process, Example 6.2.3

Example 6.2.4. *Find the zeros of*

$$P(x) = x^5 + 8x^4 + 15x^3 + 15x^2 + 8x + 1.$$

Solution: By the theorem above, we see that if r is a zero of $P(x)$, then so is $1/r$. Since P has 5 zeros, which is an odd number, P has at least one zero which is its own reciprocal! That is, either 1 or -1 is a zero. We quickly see that -1 is a zero and 1 is not, so we first divide P by $(x+1)$.

```
-1 | 1   8   15   15   8   1
   |    -1   -7   -8  -7  -1
   ------------------------------
     1   7    8    7   1 | 0
```

The other zeros of $P(x)$ are the zeros of

$$Q(x) = x^4 + 7x^3 + 6x^2 + 7x + 1.$$

Notice

$$\begin{aligned} x^{-2}Q(x) &= x^2 + 7x + 8 + 7x^{-1} + x^{-2} \\ &= (x+x^{-1})^2 + 7(x+x^{-1}) + 6. \end{aligned}$$

This shows that $x+x^{-1}$ is a zero of $z^2 + 7z + 6 = (z+1)(z+6)$. The remaining zeros of $P(x)$ satisfy the "quadratic" equations

$$x + x^{-1} = -1; \quad x + x^{-1} = -6.$$

Thus the zeros of $P(x)$ are

$$-1, \quad -\frac{1}{2} + \frac{\sqrt{3}}{2}i, \quad -\frac{1}{2} - \frac{\sqrt{3}}{2}i, \quad -3 + 2\sqrt{2}, \quad -3 - 2\sqrt{2}.$$

∎

6. Relationships Between the Zeros and the Coefficients

Exercises 6.2:

1. Prove that the rule for for "multiplying the zeros by a constant" is correct.

2. State and prove a similar rule for dividing the zeros by a non-zero constant.

3. Prove the rule for "inverting the zeros" is correct.

4. (**) Prove that Horner's process works.

5. Let $P(x) = x^3 + 4x^2 + 2x + 1$ have the zeros r_1, r_2, r_3. Find the polynomial whose zeros are as follows.

   a) $r_1 + 1$, $r_2 + 1$, $r_3 + 1$
   b) $r_1 + 5$, $r_2 + 5$, $r_3 + 5$
   c) $r_1 - 2$, $r_2 - 2$, $r_3 - 2$
   d) $-r_1 - 3$, $-r_2 - 3$, $-r_3 - 3$
   e) $2r_1$, $2r_2$, $2r_3$
   f) $5r_1$, $5r_2$, $5r_3$
   g) $-3r_1$, $-3r_2$, $-3r_3$
   h) $1/r_1$, $1/r_2$, $1/r_3$
   i) $3/r_1$, $3/r_2$, $3/r_3$
   j) $-2/r_1$, $-2/r_2$, $-2/r_3$
   k) $-r_1 + 1$, $-r_2 + 1$, $-r_3 + 1$
   l) $2r_1 + 1$, $2r_2 + 1$, $2r_3 + 1$
   m) $3r_1 - 2$, $3r_2 - 2$, $3r_3 - 2$
   n) $(r_1 - 1)/2$, $(r_2 - 1)/2$, $(r_3 - 1)/2$
   o) $2 - r_1/3$, $2 - r_2/3$, $2 - r_3/3$
   p) $(3r_1 - 1)/2$, $(3r_2 - 1)/2$, $(3r_3 - 1)/2$
   q) $(2 - r_1)^{-1}$, $(2 - r_2)^{-1}$, $(2 - r_3)^{-1}$
   r) $1 + 2/r_1$, $1 + 2/r_2$, $1 + 2/r_3$
   s) $(2 - 3r_1)^{-1}$, $(2 - 3r_2)^{-1}$, $(2 - 3r_3)^{-1}$
   t) $2 + 3/r_1$, $2 + 3/r_2$, $2 + 3/r_3$

6. Repeat the previous problem with $P(x) = 2x^3 + 4x + 7$.

7. (*) Let $P(x) = a_0 x^n + a_1 x^{n-1} + \ldots + a_0$. Prove that the sum of the products of the reciprocals of the zeros of $P(x)$ taken m at a time is $(-1)^{n-m} a_{n-m}/a_n$. Hint: Use the rule for inverting the zeros and Theorem 6.1.1.

8. For each of the following polynomials $P(x)$, find a number c and a polynomial $Q(x)$ such that $Q(x) = P(x - c)$, and the sum of the zeros of $Q(x)$ is zero.

a) $x^3 + 6x^2 - 2x + 1$

b) $x^4 - 4x^3 + 6x^2 - 4x + 1$

c) $2x^3 + 3x^2 - 9x$

d) $60x^6 + 12x^5$

e) $(x+2)^n$

f) $ax^3 + bx^2 + cx + d$

9. Let $P(x) = a_0 x^n + a_1 x^{n-1} + \ldots + a_n$. What should the real number c be if the sum of the zeros of $P(x-c)$ is zero? (See the previous problem.)

10. (*) Let $P(x) = a_0 x^n + a_1 x^{n-1} + \ldots + a_n$. If a_n is not zero, then what should the real number c be if the sum of the reciprocals of the zeros of $P(x-c)$ is zero?

6.3 Transforming the Zeros of Polynomials II

Theorem 6.3.1 (Squaring the Zeros). *Given a polynomial $P(x)$, define $Q(x)$ by*

$$Q(x^2) = P(x)P(-x).$$

Then $Q(x)$ is a polynomial, and the zeros of $Q(x)$ are the squares of the zeros of $P(x)$.

Proof. Let the zeros of $P(x)$ be $r_1, \ldots, r_n$ then

$$P(x) = a_0(x - r_1)(x - r_2) \ldots (x - r_n),$$

so $Q(x^2) = P(x)P(-x)$ is

$$\begin{aligned} Q(x^2) &= a_0^2(x - r_1)(-x - r_1)(x - r_2)(-x - r_2) \ldots (x - r_n)(-x - r_n) \\ &= (-1)^n a_0^2 (x^2 - r_1^2)(x^2 - r_2^2) \ldots (x^2 - r_n^2). \end{aligned}$$

Finally, replacing x^2 with x,

$$Q(x) = a_0^2 (-1)^n (x - r_1^2)(x - r_2^2) \ldots (x - r_n^2).$$

Example 6.3.1. *Let $r_1, \ldots, r_5$ be the zeros of*

$$P(x) = 4x^5 + 3x^4 - 2x^3 + 7x^2 - 2.$$

Find a polynomial $Q(x)$ with zeros $\left(\dfrac{1}{r_1}\right)^2, \left(\dfrac{1}{r_2}\right)^2, \ldots, \left(\dfrac{1}{r_5}\right)^2$.

6. RELATIONSHIPS BETWEEN THE ZEROS AND THE COEFFICIENTS

Solution: Begin by finding a polynomial $Q_1(x)$ whose zeros are the squares of the zeros of $P(x)$, that is,

$$\begin{aligned} Q_1(x^2) &= \left(4x^5 + 3x^4 - 2x^3 + 7x^2 - 2\right)\left(-4x^5 + 3x^4 + 2x^3 + 7x^2 - 2\right) \\ &= 16x^{10} + 25x^8 + 38x^6 + 37x^4 - 28x^2 - 4. \end{aligned}$$

So (replacing x^2 with x),

$$Q_1(x) = 16x^5 + 25x^4 + 38x^3 + 37x^2 - 28x - 4.$$

The desired polynomial $Q(x)$ is the polynomial whose zeros are the reciprocals of the zeros of $Q_1(x)$; that is,

$$Q(x) = -4x^5 - 28x^4 + 37x^3 + 38x^2 + 25x + 16.$$

■

Another common type of transformation is a translation of the polynomial $P(x)$ to find a polynomial $Q(x)$ whose coefficients have certain properties. We often do this to simplify the process of solving polynomials (see, for example, Theorems 7.2.1 and 7.3.1).

Example 6.3.2. *Let*

$$P(x) = x^4 - 4x^3 + 5x^2 - 3x + 2.$$

Translate $P(x)$ so that the coefficient of x^3 is zero.

Solution: Let r_1, r_2, r_3, r_4 be the zeros of $P(x)$, and let $r_1 + c, r_2 + c, r_3 + c, r_4 + c$ be the zeros of $Q(x)$, where $Q(x)$ has the form

$$Q(x) = x^4 + b_1 x^3 + b_2 x^2 + b_3 x + b_4.$$

We are asked to find number c such that $b_1 = 0$. Since $-b_1$ is the sum of the zeros of $Q(x)$, and 4 is the sum of the zeros of $P(x)$, we have $-b_1 = 4 + 4c$. For b_1 to be zero, we need $c = -1$, that is, we need to subtract 1 from each zero of $P(x)$. We choose to do this using Horner's process (Figure 6.3). This shows that the desired polynomial is

$$Q(x) = x^4 - x^2 - x + 1.$$

■

We will use the technique of the previous exercise enough times in this text to make it worth stating as a theorem.

6.3. Transforming the Zeros of Polynomials II

$$\begin{array}{r|rrrrr}
1 & 1 & -4 & 5 & -3 & 2 \\
  &   & 1 & -3 & 2 & -1 \\
\cline{2-6}
  & 1 & -3 & 2 & -1 & 1 \\
  &   & 1 & -2 & 0 & \\
\cline{2-5}
  & 1 & -2 & 0 & -1 & \\
  &   & 1 & -1 & & \\
\cline{2-4}
  & 1 & -1 & -1 & & \\
  &   & 1 &   & & \\
\cline{2-3}
  & 1 & 0 &   & & \\
\end{array}$$

Figure 6.3: Horner's Process, Example 6.3.2

Theorem 6.3.2. *Let $r_1, r_2, \ldots, r_n$ be the zeros of*

$$P(x) = a_0 x^n + a_1 x^{n-1} + \cdots + a_n.$$

Then $r_1 + \dfrac{a_1}{na_0}$, $r_2 + \dfrac{a_1}{na_0}$, $\ldots$, $r_n + \dfrac{a_1}{na_0}$ are the zeros of the polynomial

$$P\left(x - \dfrac{a_1}{na_0}\right) = b_0 x^n + b_1 x^{n-1} + \cdots + b_n$$

where $b_1 = 0$.

Exercises 6.3:

1. For each of the following polynomials $P(x)$, find a polynomial whose zeros are the squares of the zeros of $P(x)$.

    a) $x^2 + 1$
    b) $2x^2 + 3x + 1$
    c) $x^2 + x + 1$
    d) $x^4 - 1$
    e) $x^4 + x^3 + x^2 + x + 1$
    f) $7x^3 + 3x^2 - x + 8$

2. Let $P(x) = ax^2 + bx + c$. Find a polynomial whose zeros are the squares of the zeros of $P(x)$.

3. (*) Let $P(x) = ax^2 + bx + c$. Find a polynomial whose zeros are the fourth power of the zeros of $P(x)$. Hint: $2 \cdot 2 = 4$.

4. (*) Prove the following.

> **Theorem 6.3.3.** Let $\zeta_1, \zeta_2, \zeta_3$ be the cube roots of unity. If $P(x)$ and $Q(x)$ are such that
> $$Q(x^3) = P(\zeta_1 x)P(\zeta_2 x)P(\zeta_3 x),$$
> then $Q(x)$ is a polynomial, and the zeros of $Q(x)$ are the cubes of the zeros of $P(x)$.

5. (**) Generalize the procedures given above for squaring and cubing the zeros by proving the following.

> **Theorem 6.3.4.** Let $\zeta_1, \ldots, \zeta_n$ be the nth zeros of unity. If $P(x), Q(x)$ are defined by
> $$Q(x^n) = P(\zeta_1 x)P(\zeta_2 x)\ldots P(\zeta_n x),$$
> then $Q(x)$ is a polynomial, and the zeros of $Q(x)$ are the zeros of $P(x)$ raised to the nth power.

6. (*) Prove Theorem 6.3.2.

7. For each of the following polynomials $P(x)$, find a polynomial $Q(x)$ and a constant r such that the zeros of $Q(x)$ are r more than the zeros of $P(x)$, and the sum of the zeros of $Q(x)$ is zero.

    a) $P(x) = x^3 - 3x^2 + 2x + 7$
    b) $P(x) = 5x^3 - 4x^2 + 12x + 17$
    c) $P(x) = x^3 + bx^2 + cx + d$
    d) $P(x) = x^4 + 8x^3 - 32x^2 + 7x + 9$
    e) $P(x) = 4x^4 + 3x^3 + 21x^2 - 17x - 16$

8. (*) If the zeros of $P(x)$ are non-zero, *explain* how we may translate $P(x)$ to find a polynomial $Q(x)$ for which the sum of the inverses of the zeros is zero.

9. Let $P(x) = a_0 x^n + a_1 x^{n-1} + \ldots + a_n$ and let $Q(x) = b_0 x^m + b_1 x^{m-1} + \ldots + b_m$. Find the sum of the product of the combined list of zeros two at a time. (I write list, not set, because multiple zeros should be repeated according to their multiplicities as usual.) This is a sum of $\binom{n+m}{2}$ terms. Hint: Only requires what we learned in the first section of this chapter.

6.4 Symmetric Functions

A polynomial in several variables is called a **symmetric polynomial** or a **symmetric function** if it is not changed by any rearrangement of the variables.

6.4. Symmetric Functions

$$x^3 + y^3 + 3x^2y^2 \qquad \text{variables } x, y$$
$$3a^4 + 3b^4 + 3c^4 + 4abc(a+b+c) \qquad \text{variables } a, b, c$$
$$12(x_1x_2 + x_3x_4)(x_1x_3 + x_2x_4)(x_1x_4 + x_2x_3) \qquad \text{variables } x_1, x_2, x_3, x_4$$

The most important set of symmetric functions are the **elementary symmetric functions** on the n variables $x_1, x_2, \ldots, x_n$:

$$\begin{aligned} f_1 &= x_1 + x_2 + x_3 + \ldots + x_n \\ f_2 &= x_1x_2 + x_1x_3 + \ldots + x_{n-1}x_n \\ f_3 &= x_1x_2x_3 + x_1x_2x_4 + \ldots + x_{n-2}x_{n-1}x_n \\ \vdots &= \vdots \\ f_n &= x_1x_2x_3 \ldots x_n \end{aligned}$$

and
$$f_m = 0 \qquad \text{for all} \qquad m > n.$$

It has already been shown that if $x_1, x_2, \ldots, x_n$ are the zeros of the polynomial

$$P(x) = a_0x^n + a_1x^{n-1} + a_2x^{n-2} + \ldots + a_n$$

then coefficients of P and the elementary symmetric functions are related as follows:

$$f_1 = -\frac{a_1}{a_0}, f_2 = \frac{a_2}{a_0}, f_3 = -\frac{a_3}{a_0}, \ldots, f_n = (-1)^n\frac{a_n}{a_0}.$$

The most important result about symmetric functions is the following.

Theorem 6.4.1 (Fundamental Theorem of Symmetric Functions). *Every symmetric function in the variables $x_1, x_2, \ldots, x_n$ can be expressed as a polynomial in the elementary symmetric functions.*

Gauss proved this with a very practical algorithm. Rather than give a formal proof, we will share his algorithm by doing a couple of examples.

Example 6.4.1. *Express $P(x,y,z) = yx^3 + zx^3 + xy^3 + zy^3 + xz^3 + yz^3$ as a sum of elementary symmetric functions.*

Solution: The key is to assign an order to the terms by ordering the variables lexicographically (e.g., alphabetically), and then ordering the terms like words. For example we might choose the usual order x, then y, then z. With this ordering $x^2y^2z^2$ would come before x^2z^2, and $x^2y^3z^2 = xxyyyzz$ is before $z^2y^2x^2 = xxyyzz$.

Gauss' algorithm states that if $x^ay^bz^c$ is the first term lexicographically, then subtract the product of elementary symmetric functions: $f_1^{a-b}f_2^{b-c}f_3^c$. When expanded,

6. Relationships Between the Zeros and the Coefficients

this product begins with the same term, and each of those terms has the same length as $x^a y^b z^c$, namely $a+b+c = (a-b)+(b-c)+c$. So after we subtract $f_1^{a-b} f_2^{b-c} f_3^c$, we will have a new symmetric expression whose first term is later lexicographically and whose longest term is no longer than before we subtracted.

Now we this repeat this process with the new "first term." Because the first terms are getting later and later lexicographically (but not getting longer), the process will end with no more terms. When that happens, the original expression will equal the sum of the products of elementary functions we subtracted. Hopefully applying this with our present example will make the process clear.

For our example, the first term is $x^3 y = xxxy$, so we subtract
$$f_1^{3-1} f_2^{1-0} f_3^0 = (x+y+z)^2(xy+yz+zx),$$
which leaves
$$-2x^2y^2 - 5x^2yz - 2x^2z^2 - 5xy^2z - 5xyz^2 - 2y^2z^2.$$
Now the first term (lexicographically) is $-2x^2y^2$. Notice that we ignore the constants when determining the order of the coefficients, but not when deciding what to subtract, so we subtract $-2f_1^{2-2} f_2^{2-0} f_3^0 = -2(xy+yz+zx)^2$ which leaves
$$-x^2yz - xy^2z - xyz^2.$$
Finally, the first term is $-x^2yz$, so we subtract $-f_1^{2-1} f_2^{1-1} f_3^1 = -(x+y+z)(xyz)$, leaving zero. Looking backwards we see this means
$$P(x,y,z) = f_1^2 f_2^1 - 2f_2^2 - f_1 f_3.$$
∎

Example 6.4.2. *Express* $S = 12(x_1x_2 + x_3x_4)(x_1x_3 + x_2x_4)(x_1x_4 + x_2x_3)$ *as a sum of elementary symmetric functions.*

Solution: First, let's divide by 12, and then once finished with Gauss' algorithm, multiply by 12.

The natural order for the variables seems to be x_1, x_2, x_3 and then x_4. The first term (when expanded) is then $x_1^3 x_2 x_3 x_4$, so we subtract
$$f_1^{3-1} f_2^{1-1} f_3^{1-1} f_4^1 = (x_1 + x_2 + x_3 + x_4)^2 (x_1 x_2 x_3 x_4)$$
and get
$$x_1^2 x_2^2 x_3^2 - 2x_1^2 x_2^2 x_3 x_4 + x_1^2 x_2^2 x_4^2 - 2x_1^2 x_2 x_3^2 x_4 - 2x_1^2 x_2 x_3 x_4^2 + x_1^2 x_3^2 x_4^2 - 2x_1 x_2^2 x_3^2 x_4 - \ldots$$
Now the first term is $x_1^2 x_2^2 x_3^2$, so we subtract
$$f_1^{2-2} f_2^{2-2} f_3^{2-0} f_4^0 = (x_1 x_2 x_3 + x_1 x_2 x_4 + x_1 x_3 x_4 + x_2 x_3 x_4)^2.$$
After one more step (and returning the factor of 12), we find
$$S = 12(f_1^2 f_4 - 4 f_2 f_4 + f_3^2).$$
∎

6.4. Symmetric Functions

Using Computer Algebra Systems

Gauss' algorithm is quite straight forward, so it can be programmed easily. In MapleMaple, the command is `convert(...,elsymfun)`. For example, to express

$$(x^3 + y + z)(y^3 + x + z)(z^3 + x + y)$$

as a sum of elementary symmetric functions we need only enter

`convert((x^3+y+z)*(y^3+x+z)*(z^3+x+y),elsymfun)`

which returns

$$(x+y+z)^5 - 4(xy+xz+yz)(x+y+z)^3 - 3(xy+xz+yz)(xyz)(x+y+z)^2$$
$$+ (xy+xz+yz)^3(x+y+z) + 5xyz(x+y+z)^2 + 2(xy+xz+yz)^2(x+y+z)$$
$$+ 5(x+y+z)(x^2y^2z^2) - (xy+xz+yz)^2xyz + x^3y^3z^3$$
$$+ (xy+xz+yz)(x+y+z) - 2(xy+xz+yz)xyz - xyz.$$

This is quite a mouthful! After some rearranging we can write this as

$$f_1^5 - 4f_1^3 f_2 - f_1^2(3f_2 f_3 - 5f_3) + f_1(f_2^3 + 2f_2^2 + f_2 + 5f_3^2) - (2f_2 + f_2^2 + 1 - f_3^2)f_3.$$

Exercises 6.4:

1. Show that the following definition is equivalent to the definition of symmetric polynomial given in the text. (Two definitions are equivalent if whenever an object satisfies either of the definitions, then it satisfies both.)

   Definition: A polynomial in several variables is a symmetric polynomial if it is not changed when any two of the variables are interchanged.

2. Determine if the following are symmetric functions.

   a) $(x_1 - x_2)^2$

   b) $x_1^2 + x_1 x_2 + x_2^2$

   c) $(x_1 + x_2)(x_1 + x_3)(x_2 + x_3)$

   d) $(x_1 - x_2)^2 (x_1 - x_3)^2 (x_2 - x_3)^2$

   e) $(x_1 x_2 - x_3 x_4)(x_1 x_3 - x_2 x_4)(x_1 x_4 - x_2 x_3)$

   f) $(x_1 + x_2 + x_3 - x_4)(x_1 + x_2 - x_3 - x_4)(x_1 - x_2 - x_3 - x_4)$

   g) $(x_1 + x_2 + x_3 + x_4)^2 + (x_1 + x_2 + x_3 + x_5)^2 + (x_1 + x_2 + x_4 + x_5)^2$

   h) $(x_1 + x_2)^2 x_1 + (x_3 + x_3)^2 x_1 x_2 + (x_2 + x_3)^2 x_1 x_2 x_3$

   i) $(x_1 - x_2)^2 x_3 + (x_1 - x_3)^2 x_2 + (x_2 - x_3)^2 x_1$

   j) $(x_1 - x_2)^2(x_1 - x_3)^2 + (x_1 - x_2)^2(x_2 - x_3)^2 + (x_1 - x_3)^2(x_2 - x_3)^2$

6. Relationships Between the Zeros and the Coefficients

3. Express the following in terms of the elementary symmetric functions.

   a) $(x_1 - x_2)^2$

   b) $x_1^2 + x_1 x_2 + x_2^2$

   c) $(x_1 x_2 + x_1 x_3)(x_1 x_2 + x_2 x_3)(x_2 x_3 + x_1 x_3)$

   d) $(x_1 + x_2 - x_3 - x_4)(x_1 - x_2 + x_3 - x_4)(x_1 - x_2 - x_3 + x_4)$

   e) $(x_1 - x_2)^2 (x_1 - x_3)^2 (x_2 - x_3)^2$

4. Let x_1, x_2 be the zeros of $ax^2 + bx + c$. Write $(x_1 - x_2)^2$ first in terms of the elementary symmetric functions f_1, f_2; and then in terms of a, b, c.

5. Find the simple condition on the coefficients b and c of the polynomial $P(x) = x^3 + bx + c$ under which one zero is the negative of another zero. Hint: Consider the symmetric function $(r_1 + r_2)(r_2 + r_3)(r_3 + r_1)$ which is zero when ...

6. Find the simple condition on the coefficients a, b and c of the polynomial $P(x) = x^4 + ax^2 + bx + c$ under which one zero is the negative of another zero. Hint: Consider the symmetric function $(r_1 + r_2)(r_2 + r_3)(r_3 + r_4)(r_4 + r_1)(r_1 + r_3)(r_2 + r_4)$ which is zero when ...

7. Use Maple to verify the following conditions on the coefficients of $x^3 + ax^2 + bx + c$ under which:

   a) two zeros are equal $(4a^3 c - a^2 b^2 - 18abc + 4b^3 + 27c^2 = 0)$;

   b) one zero is twice another $(36a^3 c - 8a^2 b^2 - 182abc + 36b^3 + 343c^2 = 0)$;

   c) one zero is the negative one half of another $(4a^3 c + 8a^2 b^2 - 18abc + 4b^3 + 27c^2 = 0)$.

   Hint: Consider the symmetric functions

   a) $(r_1 - r_2)^2 (r_2 - r_3)^2 (r_3 - r_1)^2$,

   b) $(r_1 - 2r_2)(r_2 - 2r_1)(r_1 - 2r_3)(r_3 - 2r_1)(r_3 - 2r_1)(r_1 - 2r_3)$ and

   c) $(r_1 + 2r_2)(r_2 + 2r_1)(r_1 + 2r_3)(r_3 + 2r_1)(r_3 + 2r_1)(r_1 + 2r_3)$ respectively.

8. (*) Let $P(x) = a_0 x^n + a_1 x^{n-1} + \ldots + a_n$. Show that there is a polynomial in the variables $a_0, a_1, \ldots, a_n$ which is zero if and only if $P(x)$ has a repeated zero. Hint: Consider the following polynomial in $r_1, r_2, \ldots, r_n$:

$$(r_1 - r_2)^2 (r_1 - r_3)^2 \cdots (r_{n-1} - r_n)^2.$$

   Note that you are showing the polynomial exists, not finding it! (It will have 59 terms if $n = 5$ and 246 if $n = 6$.)

9. Express the following ratios of elementary symmetric functions in terms of $x_1, x_2, \ldots, x_n$:

$$\frac{f_{n-1}}{f_n}, \quad \frac{f_{n-2}}{f_n}, \quad \ldots, \quad \frac{f_1}{f_n}, \quad \frac{1}{f_n}.$$

10. (*) Use Gauss' algorithm to prove Theorem 6.4.1. Hint: Just formalize the algorithm presented in the examples.

6.5 Newton's Identities

In this section we look at a very special case: writing the sums of powers of the zeros of a polynomial in terms of its coefficients.

Example 6.5.1. *Let $S_m = x_1^m + x_2^m + \ldots + x_n^m$. Express S_m for $m = 1, 2, 3$ and 4 as polynomials in the elementary symmetric functions.*

Solution: Trivially $S_1 = f_1$. For $n = 2$ we see
$$x_1^2 + x_1^2 + \ldots + x_1^2 = (x_1 + x_2 + \ldots + x_n) - 2(x_1x_2 + x_1x_3 + \ldots + x_{n-1}x_n)$$
which is $S_2 = f_1^2 - 2f_2$. We may easily show
$$S_3 = f_1^3 - 3f_1f_2 + 3f_3$$
$$S_4 = f_1^4 - 3f_1^2f_2 + 4f_1f_3 - 4f_4 + 2f_2^2$$

∎

We may also calculate S_n recursively by writing these identities in the following form, which are known as Newton's Identities
$$S_k - f_1 S_{k-1} + f_2 S_{k-2} - f_3 S_{k-3} + \ldots + k(-1)^k f_k = 0.$$

Our focus in this text is polynomials, so we restate these identities in terms of the coefficients below. We also include the relevant definitions to make it self contained.

Theorem 6.5.1 (Newton's Identities). *Let*
$$P(x) = x^n + a_1 x^{n-1} + a_2 x^{n-2} + \ldots + a_{n-1}x + a_n$$
be a polynomial with (not necessarily distinct) zeros $x_1, x_2, \ldots, x_n$. For each positive integer m let $S_m = x_1^m + x_2^m + \ldots + x_n^m$. Define $a_k = 0$ for all integers $k > n$. Then for all $k > 0$,
$$S_k + a_1 S_{k-1} + a_2 S_{k-2} + a_3 S_{k-3} + \ldots + k f_k = 0.$$

6. Relationships Between the Zeros and the Coefficients

In particular, this theorem yields the following for monic polynomials.

$$S_1 + a_1 = 0$$
$$S_2 + a_1 S_1 + 2a_2 = 0$$
$$S_3 + a_1 S_2 + a_2 S_1 + 3a_3 = 0$$
$$S_4 + a_1 S_3 + a_2 S_2 + a_3 S_1 + 4a_4 = 0$$

For a sketch of the proof, see exercise 7.

Example 6.5.2. *Let $P(x) = x^3 - 3x^2 + 2x + 1$. Find the sum of the seventh powers of the zeros of P.*

Solution: Remember that the coefficients of monic polynomials are defined by the elementary symmetric functions of its zeros; here $P(x) = x^3 - f_1 x^2 + f_2 x + -f_3$. So in this example we are given that $f_1 = 3$, $f_2 = 2$, $f_3 = -1$ and $f_n = 0$ for all $n > 3$. Using Newton's Identities we find recursively:

$$S_1 = 3$$
$$S_2 = 3(3) - 2(2) = 5$$
$$S_3 = 3(5) - 2(3) + 3(-1) = 6$$
$$S_4 = 3(6) - 2(5) - 1(3) = 5$$
$$S_5 = 3(5) - 2(6) - 1(5) = -2$$
$$S_6 = 3(-2) - 2(5) - 1(6) = -22$$
$$S_7 = 3(-22) - 2(-2) - 1(5) = -67$$

So -67 is the desired sum. ■ By the way, in that last example $S_6 = -22 < 0$. Since S_6 is the sum of the sixth powers of the zeros of $P(x)$, if the zeros were are real, then we'd have $S_6 > 0$. This shows that this polynomial has at least one complex zero. The coefficients are real, so the these come in conjugate pairs. This shows our cubic polynomial has exactly two complex zeros.

Using Computer Algebra Systems

Though we leave the proof of Newton's Identities to Exercise 7, it is useful to single out the following key result:

$$x \frac{P'(x)}{P(x)} = S_0 + \frac{S_1}{x} + \frac{S_2}{x^2} + \frac{S_3}{x^3} \cdots . \tag{6.1}$$

Computer Algebra Systems such as `Maple` have this this type of expansion built in. For example, the `Maple` commands

6.5. Newton's Identities

```
p:= x^3-3*x^2+2*x+1;
asympt(x*diff(p,x)/p, x, 9);
```

generate

$$3 + 3x^{-1} + 5x^{-2} + 6x^{-3} + 5x^{-4} - 2x^{-5} - 22x^{-6} - 67x^{-7} - 155x^{-8} + \mathrm{O}\left(x^{-9}\right).$$

Once again we see the sum of the seventh powers of the zeros from example 6.5.2 is -67 (and we have also verified each of the smaller powers). The last term, $O(x^{-9})$, tells us that the rest of this infinite expansion involves powers of x of nine or higher in the denominators. This is called "big-Oh" notation.

Newton's Identities can also be extend to negative powers. Again, let

$$P(z) = a_0 x^n + a_1 x^{n-1} + \cdots + a_n$$

be a polynomial with zeros (counted with multiplicity) $x_1, x_2, \ldots, x_n$. We assume that $p_n = 0$, and

$$S_{-k} = x_1^{-k} + x_2^{-k} + \ldots + x_n^{-k} = \frac{1}{x_1^k} + \frac{1}{x_2^k} + \ldots + \frac{1}{x_n^k}.$$

Then for all $j \leq n$ we have

$$(n-j)a_j + a_{j+1} S_{-1} + \cdots + a_n S_{j-n} = 0. \tag{6.2}$$

The proof involves the following formal power series result.

$$-x \frac{P'(x)}{P(x)} = S_{-1} x + S_{-2} x^2 + S_{-3} x^3 + \ldots$$

This can also be easily evaluated via Maple

```
series(-z*diff(p,z)/p,z,9);
```

For the example above, this yields

$$-2x + 10x^2 - 29x^3 + 90x^4 - 277x^5 + 853x^6 - 2627x^7 + 8090x^8 + \mathrm{O}\left(x^9\right)$$

Exercises 6.5:

1. Find the sum of the mth powers of the zeros of the given polynomials.

    a) $x^2 + x + 1$, $m = 3$
    b) $x^2 + x + 1$, $m = 4$
    c) $x^3 + 3$, $m = 3$
    d) $x^3 + 3$, $m = 9$

6. Relationships Between the Zeros and the Coefficients

    e) $x^4 + 2x^3 + 3x^2 + x + 7$, $m = 5$

    f) $x^4 - x^3 + 2x^2 + x - 1$, $m = 5$

    g) $x^4 + x^3 + x^2 + x + 1$, $m = 5$

    h) $x^5 + 3x^3 + x$, $m = 8$

    i) $x^6 + 5x^5 - 3x^4 + 2x^3 + 6x^2 + 4x - 8$; $M = 8$

2. Let a, b, c be real numbers for which $a + b + c = 3$, $a^2 + b^2 + c^2 = 5$, and $a^3 + b^3 + c^3 = 7$, find $a^4 + b^4 + c^4$. Hint: Let $P(x) = (x-a)(x-b)(x-c)$.

3. Let $P(x) = d_1 x^8 + d_2 x^7 + d_3 x^6 + \ldots + d_8 x + d_9$, where $d_1 d_2 d_3 d_4 d_5 d_6 d_7 d_8 d_9$ is your student ID number (change the last digit to 1 if it is 0). Use Maple to find the sum of the nth powers of the zeros for n from -9 to 9.

4. Given the polynomial $p(x) = x^4 + x^3 + x^2 + x + 1$, find $S_0, S_1, S_2, \ldots, S_{20}$. They have a simple pattern; prove that pattern holds. Hint: First multiply $p(x)$ by $(x-1)$ (this should tell you exactly what the four zeros are). Next settle the case where n is a multiple of five; this should be very easy. Now do the other cases. If you are stuck, graph the four zeros, then graph their squares (cubes, fourth powers).

5. (*) Prove that it is possible to express S_n ($n < 0$) in terms of the elementary symmetric functions (you need not find the explicit representation).

6. Write a program that accepts as input the coefficients of a polynomial together with an integer m, and outputs S_m.

7. (**) Prove Newton's Identities. Hint: Write
$$P(x) = x^n + a_1 x^{n-1} + a_2 x^{n-2} + \cdots + a_n = (x - r_1)(x - r_2) \cdots (x - r_n)$$

Take the logarithm, and then differentiate, so:
$$\frac{P'(x)}{P(x)} = \frac{1}{x - r_1} + \frac{1}{x - r_2} + \cdots + \frac{1}{x - r_n}.$$

Multiply by x, expand each term as a geometric series and regroup to get
$$x \frac{P'(x)}{P(x)} = S_0 + \frac{S_1}{x} + \frac{S_2}{x^2} + \frac{S_3}{x^3} \cdots.$$

Now multiply by $P(x)$ and compare with
$$xP'(x) = nx^n + (n-1)a_1 x^{n-1} + \cdots + a_{n-1} x$$

and (with clever manipulation), you are done.

The Algebraic Solution of Polynomials of Low Degree

The study of mathematics develops and sets into operation a mental organism more valuable than a thousand eyes, because through it alone can truth be apprehended.

Plato, *Republic*

One cannot escape the feeling that these mathematical formulas have an independent existence and an intelligence of their own, that they are wiser than we are, wiser even than their discoverers, that we get more out of them than was originally put into them.

Heinrich Hertz—discoverer of radio waves

Chapter 7

Algebraic Solutions of Polynomials of Low Degree

In this chapter we present algebraic solutions to the general equations of degree one, two, three and four. There are no algebraic solutions to the general equations of degree five or higher.

7.1 General Polynomial Equations of Degree n

WHAT IS A SOLUTION to a polynomial equation? It should be a way to express the zeros exactly in terms of what we are given—the coefficients of the polynomial. A solution should be finite. Using numerical methods we can approximate a solution, but we usually cannot determine solutions exactly in a finite number of steps with a limited set of elementary operations. This leads to the following definition.

An **algebraic solution** to a polynomial equation (for example the general equation of degree n) is a finite algebraic expression for its zeros written in terms of the coefficients and using only the elementary operations of algebra: addition, subtraction, multiplication, division, and extracting m^{th} roots. An algebraic solution is also called a **solution by radicals**, and an equation with an algebraic solution is said to be **solvable by radicals**.

Example 7.1.1. *The general equation of degree one, called the general linear equation, is*

$$ax + b = 0$$

(where $a \neq 0$) and has the algebraic solution

$$x = -\frac{b}{a}.$$

7. Algebraic Solutions of Polynomials

Example 7.1.2. *The general equation of degree two, called the general quadratic, is*

$$ax^2 + bx + c = 0,$$

(where $a \neq 0$) and has the algebraic solution

$$x = \frac{-b \pm \sqrt{b^2 - 4ac}}{2a}.$$

This solution is called the **quadratic formula**.

(The proof of the quadratic formula is left to the reader in exercise 1 and its use is explored in the next couple problems.)

For each positive integer n, we hope for a solution that will solve any equation of degree n. For example, the quadratic formula is an algebraic solution for the general equation of degree two. We now want to learn how to find an algebraic solution to the **general equation of degree** n:

$$a_0 x^n + a_1 x^{n-1} + \ldots + a_{n-1} x + a_n = 0$$

where the coefficients are themselves are viewed as variables.

In this chapter we present solutions to the general equations of degree three and four. We remarked earlier there are no solutions to the general equations of degree five and higher. This does not mean we cannot solve particular equations or even some large classes of polynomial equations with higher degree, just that we cannot solve all of them.

Theorem 7.1.1. *There are no algebraic solutions to the general equations of degree five and higher.*

The reason this is true is that there are polynomial equations of every degree greater than four whose zeros cannot be expressed in terms of radicals. For example the zeros of

$$x^5 - x - 1$$

cannot be written using rational numbers and radical signs. The proof of this fact (and of the theorem above) belong to a branch of abstract algebra known as Galois Theory and is beyond the scope of this text. The proof is covered in the second term of most abstract algebra courses.

Exercises 7.1:

1. Prove that the quadratic formula is correct. Hint: Complete the square.

2. The quadratic formula may fail on your calculator if b^2 is far larger than $4ac$, in which case the numerator of
$$x = \frac{-b + \sqrt{b^2 - 4ac}}{2a}$$
may round to zero. Test this with $a = c = 1$ and $b = 10^{12}$ on your calculator. Show that this same solution is given by
$$x = \frac{-2c}{b + \sqrt{b^2 - 4ac}}$$
(where now the two very close terms are added instead of subtracted), and again try $a = c = 1$ and $b = 10^{12}$ on your calculator with this formula. Hint: Multiply the numerator and denominator of the first expression by $-b - \sqrt{b^2 - 4ac}$.

3. Let $P(x) = ax^2 + bx + c$. The quantity $D = b^2 - 4ac$ is called the **discriminant** of the quadratic. Assuming that a, b and c are real show the following.

   a) If $D > 0$, then $P(x)$ has two distinct real zeros.

   b) If $D = 0$, then $P(x)$ has two equal real zeros.

   c) If $D < 0$, then $P(x)$ has two distinct complex zeros.

4. Find a and b so that the following polynomials have two common zeros: $x^3 - 6x^2 + ax - 3$ and $x^3 - x^2 + bx + 2$. Hint: Subtract then use the discriminant from the previous problem.

5. Use Theorem 6.3.2 to translate $P(x) = ax^2 + bx + c$ so that the coefficient of x is zero.

6. (*) Show that if we drop the finiteness condition in the definition of algebraic solution, then every polynomial equation has an algebraic solution.

7. Find the general solution to equations of the form $ax^n + c = 0$.

8. Find the general solution to equations of the form $ax^{2n} + bx^n + c = 0$.

9. Explain why a solution to the previous problem does not contradict the theorem of this section.

10. (*) Find the general solution to equations of the form
$$ax^4 + bx^3 + cx^2 + bx + a = 0.$$
Hint: Divide by x^2. Let $y = x + x^{-1}$. What is $y^2 - 2$?

11. (**) Find the general solution to equations of the form
$$ax^5 + bx^4 + cx^3 + cx^2 + bx + a = 0.$$
Hint: Divide by $(x+1)$, then use the hint for the previous problem.

7. Algebraic Solutions of Polynomials

7.2 Solution of the General Cubic

In this section we solve the general cubic equation:

$$P(x) = ax^3 + bx^2 + cx + d = 0.$$

We first simplify the polynomial itself. By adding $b/3a$ to each zero of $P(x)$ (Theorem 6.3.2), we get a new polynomial without an x^2 term (one whose x^2 coefficient is zero). We then divide this polynomial by its leading coefficient to get a polynomial in the **reduced form** of the general cubic:

$$P_t(x) = x^3 + 3px + 2q,$$

where

$$p = \frac{3ac - b^2}{9a^2}, \qquad q = \frac{27a^2d - 9abc + 2b^3}{54a^3}.$$

Once we find the zeros of the reduced form, we can translate back to get the zeros of the original polynomial.

The term $D = -p^3 - q^2$ is called **the discriminant of the cubic**, for if p and q are real we have the following.

- If $D > 0$, then P and P_t have three real (and unequal) zeros.

- If $D = 0$, then P and P_t have three real zeros, and at least two are equal.

- If $D < 0$, then P and P_t have one real and two complex zeros.

For a proof of these three statements, see exercises 1, 2 and 3. (There are other possible definitions of the discriminant; see exercises 9–11.) The first case $D > 0$ is called the **irreducible case**, and we look it more closely in section 7.3.

For historic reasons, many older books use $p^3 + q^2$ and reverse the inequality signs, but use of the negative signs is more mathematically consistent.

Theorem 7.2.1 (Cardan's Solution to the General Cubic). *The zeros of $x^3 + 3px + 2q$ are given by*

$$r_1 = s_1 + s_2, \quad r_2 = \omega s_1 + \omega^2 s_2, \quad \text{and} \quad r_3 = \omega^2 s_1 + \omega s_2$$

where

$$s_1 = \left(-q + (p^3 + q^2)^{1/2}\right)^{1/3}, \qquad s_2 = \left(-q - (p^3 + q^2)^{1/2}\right)^{1/3}$$

and ω is a primitive cube root of one. The cube roots s_1, s_2 must be chosen so that

$$s_1 s_2 = -p.$$

7.2. Solution of the General Cubic

We derive Cardan's solution at the end of this section.

Recall that the three cube roots of one are the zeros of $x^3 - 1 = (x-1)(x^2+x+1)$, so they are

$$1, \quad \omega = \frac{-1+\sqrt{-3}}{2}, \quad \text{and} \quad \omega^2 = \frac{-1-\sqrt{-3}}{2}.$$

Once we have found the zeros r_1, r_2, and r_3 of P_t, then the zeros of the original polynomial P are

$$r_1 - \frac{b}{3a}, \quad r_2 - \frac{b}{3a}, \quad \text{and} \quad r_3 - \frac{b}{3a}.$$

The complex zeros of real cubics come in conjugate pairs, so it is sometimes useful to recall that ω^2 is the complex conjugate of ω, that is, $\omega^2 = \overline{\omega}$.

Example 7.2.1. *Find the zeros of $P(x) = x^3 - 6x + 6$.*

Solution: This polynomial is already in reduced form with $p = -2$ and $q = 3$. Since $D = -p^3 - q^2 = -1 < 0$, $P(x)$ has one real and two imaginary zeros. To find the zeros using Cardan's solution we first calculate s_1 and s_2:

$$s_1 = \left(-3 + 1^{1/2}\right)^{1/3} = -\sqrt[3]{2}$$

$$s_2 = \left(-3 - 1^{1/2}\right)^{1/3} = -\sqrt[3]{4}.$$

Notice $s_1 s_2 = -p = 2$. Using these we find that the zeros are

$$r_1 = -\sqrt[3]{2} - \sqrt[3]{4}$$
$$r_2 = -\sqrt[3]{2}\left(-1 + \sqrt{-3}\right)/2 - \sqrt[3]{4}\left(-1 - \sqrt{-3}\right)/2$$
$$r_3 = -\sqrt[3]{2}\left(-1 - \sqrt{-3}\right)/2 - \sqrt[3]{4}\left(-1 + \sqrt{-3}\right)/2.$$

Or slightly simplified,

$$-\sqrt[3]{2} - \sqrt[3]{4}, \quad \frac{\sqrt[3]{2} + \sqrt[3]{4}}{2} \pm \frac{\sqrt{3}\left(\sqrt[3]{2} - \sqrt[3]{4}\right)}{2}i.$$

Here r_3 is the complex conjugate of r_2, so that once we had found r_2, there was no need to use Cardan's solution to find r_3—this trick saves a little work. ∎

Example 7.2.2. *Find the zeros of $P(x) = x^3 - 13x - 12$.*

Solution: The polynomial is also in reduced form with $p = -13/3$, $q = -6$ and $D = 1225/27$. Since $D > 0$, this polynomial has three real zeros.

$$s_1 = \left(6 + (-1225/27)^{1/2}\right)^{1/3} = 2 + 3^{-1/2}i$$

$$s_2 = \left(6 - (-1225/27)^{1/2}\right)^{1/3} = 2 - 3^{-1/2}i$$

7. Algebraic Solutions of Polynomials

Thus we have,

$$r_1 = (2 + 3^{-1/2}i) + (2 - 3^{-1/2}i) = 4$$
$$r_2 = (2 + 3^{-1/2}i)\omega + (2 - 3^{-1/2}i)\omega^2 = -3$$
$$r_3 = (2 + 3^{-1/2}i)\omega^2 + (2 - 3^{-1/2}i)\omega = -1.$$

■

For the last example, the Rational Zero Theorem 5.1.1 would have been easier! We will end with a more extreme example of these, one that Cardan himself used.

Example 7.2.3. *Find the zeros of $P(x) = x^3 + 6x - 20$.*

Solution: Here $p = 2$, $q = -10$, and $D = -108$ (so we have one real and two complex zeros).

$$s_1 = \left(10 + 6(3)^{1/2}\right)^{1/3}, \qquad s_2 = \left(10 - 6(3)^{1/2}\right)^{1/3},$$

so we have (choosing the cube roots to be real),

$$r_1 = \sqrt[3]{10 + 6\sqrt{3}} + \sqrt[3]{10 - 6\sqrt{3}}$$
$$r_2 = \sqrt[3]{10 + 6\sqrt{3}}\,\omega + \sqrt[3]{10 - 6\sqrt{3}}\,\omega^2$$
$$r_3 = \sqrt[3]{10 + 6\sqrt{3}}\,\omega^2 + \sqrt[3]{10 - 6\sqrt{3}}\,\omega$$

■

It is easy to check that 2 is a zero of the polynomial in the previous example, so it must be that

$$2 = \sqrt[3]{10 + 6\sqrt{3}} + \sqrt[3]{10 - 6\sqrt{3}},$$

because they both satisfy $x^3 + 6x - 20$, which only has one real root. I challenge you to try to show this directly! In general, it can be very difficult to tell if two algebraic objects are equal.

Historical Note

This solution is known as **Cardan's solution**, even though it was not discovered by Cardan. A partial solution to the general cubic was first discovered by Scipione del Ferro around 1505. Scipione decided to keep his solution secret, sharing it only with his students. These students became famous by giving public demonstrations of their problem solving ability and winning the public mathematical competitions. About 40 years later, the outsider Tartaglia (actually a nickname meaning "the stammerer"), began to win the public mathematical competitions – he had broken Scipione's monopoly by discovering a solution himself! Tartaglia also decided to keep this solution secret, but later shared the technique with the important mathematician-physician Cardan (perhaps to gain favor and employment with the Spanish military). In 1545, Cardan publishing a complete solution of the cubic (including the case Tartaglia had solved), and his student Ferrari's solution to the general quartic, in a book called *Ars Magna* (literally "Great Art".) This greatly angered Tartaglia, who challenged Ferrari to a public debate in Milan in 1548. Tartaglia lost, and to this day his solution is called Cardan's solution.

Derivation of Cardan's Solution

We begin with the reduced cubic equation

$$x^3 + 3px + 2q = 0.$$

After making the substitution $x = u + v$, the reduced cubic becomes

$$u^3 + v^3 + 3(p + uv)(u + v) + 2q = 0. \tag{7.1}$$

We now have one equation in two unknowns, which is one too many degrees of freedom. So we may carefully choose a second equation relating u and v, and then hopefully solve the resulting system of two equations. A choice that makes the system solvable is

$$uv = -p. \tag{7.2}$$

Using this, Equation (7.1) becomes

$$u^3 + v^3 = -2q.$$

Cubing Equation (7.2) we find

$$u^3 v^3 = -p^3.$$

Since we know the sum and the product of u^3 and v^3, we may write down a quadratic polynomial which has them as zeros. In particular, u^3 and v^3 are the zeros of

$$s^2 + 2qs - p^3,$$

which are clearly

$$-q + \sqrt{p^3 + q^2} \quad \text{and} \quad -q - \sqrt{p^3 + q^2}.$$

Let s_1 and s_2 be the cube roots of these expressions, chosen so that $s_1 s_2 = -p$. Then the possible values of u and v are

$$u = s_1 \omega^j \quad \text{and} \quad v = s_2 \omega^k$$

7. ALGEBRAIC SOLUTIONS OF POLYNOMIALS

where j, k are in the set $\{0, 1, 2\}$. Now $uv = s_1 s_2 \omega^{j+k} = -p\omega^{j+k}$, which by Equation 7.2 must equal $-p$, so the pair (j, k) is $(0,0)$, $(1,2)$ or $(2,1)$. Thus the zeros $x = u + v$ of $P_t(x)$ must be

$$r_1 = s_1 + s_2, \qquad r_2 = s_1\omega + s_2\omega^2, \qquad \text{and} \qquad r_2 = s_1\omega^2 + s_2\omega$$

respectively.

Exercises 7.2:

1. Show that if the discriminant is zero, then the zeros of P_t are: $2(-q)^{1/3}, q^{1/3}$, and $q^{1/3}$. This shows that if $D = 0$, then P_t has a repeated zero.

2. Show that if the cubic has real coefficients and $D < 0$, then the cubic has one real and two complex zeros. Hint: When the discriminant is negative then s_1, s_2 may be chosen to be real so r_1 is real and r_2, r_3 are conjugate.

3. Show that if $D > 0$, then the cubic has three real zeros. Hint: Show that in this case s_2 (in theorem 7.2.1) may be chosen to be the complex conjugate of s_1, so r_1, r_2 and r_3 are self conjugate.

4. (*) Give a polynomial with real coefficients and zeros $r_1, r_2, \ldots, r_n$, define $D = \prod_{i \neq j} (r_i - r_j)^2$. Show that:

   a) $D > 0$ if $p(x)$ has all real distinct zeros;

   b) $D = 0$ if and only if two of the zeros of $p(x)$ are equal; and

   c) $D < 0$ if $p(x)$ has exactly two complex (non-real) zeros.

5. Find the zeros of the following polynomials.

   a) $2x^3 - 3x + 1$
   b) $2x^3 - 12x + 40$
   c) $x^3 - 6x^2 + 6x - 5$
   d) $x^3 + 15x^2 + 105x + 245$
   e) $3x^3 + 6x^2 + 2$
   f) $x^3 + 3x^2 + 51x - 47$
   h) $x^3 - 6x + 6$
   i) $x^6 - 18x^2 - 30$

6. Draw a graph dividing the plane with axes p and q (instead of x and y) into three regions: those for which $x^3 - px + q$ has (a) three distinct real zeros, (b) a pair of equal zeros and (c) only one real zero.

7. (*) Write a short essay (three pages double spaced) on one or more of the important figures involved in the solution of the cubic (Scipione, Tartaglia, Cardan or Ferrari).

8. Write a program to check to see if a cubic $P(x)$ is in the irreducible case, and *if not*, to solve it. Hint: The program need not use complex arithmetic. A solution for the irreducible case involving only real arithmetic is given in the next section.

9. Let D be the discriminant of the cubic as defined above. Show that if q is any quantity that is always positive, then D may be replaced by qD when testing whether the zeros of the cubic are real, equal or complex. (For this reason, D is often called *a* discriminant, as is qD.)

10. Show that a discriminant of $x^3 + mx + n$ is $-4m^3 - 27n^2$. Many authors define this to be "the" discriminant. Hint: This is $108D$. See problem 9.

11. Show that a discriminant of $ax^3 + bx^2 + cx + d$ is
$$-4ac^3 + b^2c^2 + 18abcd - 27a^2d^2 - 4db^3.$$

Some authors define this to be "the" discriminant. Hint: This is $432a^4 D$. See problem 9.

7.3 Solution of the Cubic: Irreducible Case

Recall that the irreducible case of the general cubic $P(x)$ is the case where the discriminant is positive ($D > 0$). In this case $P(x)$ has three distinct real zeros. The irreducible case has second solution, called the **trigonometric solution**, which is easier to use than Cardan's solution.

Theorem 7.3.1 (The Trigonometric Solution to the General Cubic). *Let $P_t(x)$ be a cubic with positive discriminant, given in the reduced form:*
$$P_t(x) = x^3 + 3px + 2q.$$

Define the angle θ by
$$\cos\theta = \frac{-q}{\sqrt{-p^3}}, \qquad (0 < \theta < \pi).$$

Then the zeros of $P(x)$ are given by
$$r_1 = 2\sqrt{-p}\cos\left(\frac{\theta}{3}\right),$$
$$r_2 = 2\sqrt{-p}\cos\left(\frac{\theta}{3} + \frac{2\pi}{3}\right),$$
and
$$r_3 = 2\sqrt{-p}\cos\left(\frac{\theta}{3} + \frac{4\pi}{3}\right).$$

7. ALGEBRAIC SOLUTIONS OF POLYNOMIALS

This solution is especially easy to implement using the arc-cosine and square root functions on hand calculators. The derivation of this solution is left to the reader (see exercises 6–8).

Example 7.3.1. *Approximate the zeros of $P(x) = x^3 - 13x - 12$.*

Solution: $P(x)$ is already in the reduced form with $p = -13/3$, $q = -6$ and the discriminate D is $1224/27 > 0$. We first approximate the angle θ using

$$\cos \theta = \frac{6}{\frac{13}{3}\sqrt{\frac{13}{3}}} = \frac{18\sqrt{3}}{13\sqrt{13}}.$$

so $\theta \approx 0.8431047045$ radians. Thus we find:

$$r_1 \approx 2\sqrt{\tfrac{13}{3}} \cos(0.2810349015) \approx 4$$

$$r_2 \approx 2\sqrt{\tfrac{13}{3}} \cos(2.375430004) \approx -3$$

$$r_3 \approx 2\sqrt{\tfrac{13}{3}} \cos(4.469825106) \approx -1$$

Having found that the zeros are approximately 4, -3 and -1, we check and see that these are exactly the zeros. ∎

Exercises 7.3:

1. Show that in the irreducible case $p < 0$, so $(-p)^{1/2}$ is a real number. Also show that either square root of $-p$ may be used as long as the same zero is used each time $(-p)^{1/2}$ appears.

2. For the trigonometric solution to work $-q/(-p^3)^{1/2}$ must between -1 and 1. State why and show that this is the case if and only if the discriminant $-p^3 - q^2$ is positive.

3. Find the zeros of the following polynomials using the trigonometric solution. First state them exactly, then approximately.

    a) $x^3 - 6x^2 + 8$
    b) $2x^3 + 3x^2 - 2x - 3$
    c) $x^3 - 3x^2 - 2x + 3$
    d) $9x^3 + x^2 - 1$
    e) $x^3 - 6x^2 + 8x + 1$
    f) $x^3 - 6x^2 + 10x - 3$

7.3. Solution of the Cubic: Irreducible Case

4. ▣ Write a program to check to see if a cubic $P(x)$ has a positive discriminant, and if so, to solve it.

5. ▣ (*) Write a program to solve the general cubic. Hint: Use the previous problem and problem 8 of the previous section.

6. Let $a = \cos(3\alpha)$, then show that $x = 2\cos(\alpha)$ satisfies
$$x^3 - 3x - 2a = 0.$$
This is known as the **trisection equation**. Hint: Use the triple angle formula for cosine.

7. Show that the zeros of the trisection equation (see the previous problem) are
$$2\cos(\alpha), \quad 2\cos(\alpha + 2\pi/3), \quad \text{and} \quad 2\cos(\alpha + 4\pi/3).$$
Hint: Compare 3α, $3(\alpha + 2\pi/3)$, and $3(\alpha + 4\pi/3)$.

8. (*) Verify that the trigonometric solution is correct. Hint: Use the two previous problems as models and set $x = 2(-p)^{1/2}\cos(\alpha)$ and $q = -(-p)^{3/2}\cos(3\alpha)$.

9. (**) Show that the trigonometric solution is actually a solution to the general cubic if we allow θ to be a complex number.

10. (**) Derive Cardan's solution from the trigonometric solution using the identity $\cos(\theta) = (e^{i\theta} + e^{-i\theta})/2$. Hint: Show the following.
    a) $e^{\pm i\theta} = a \pm (a^2 - 1)^{1/2}$.
    b) $(-p)^{1/2} e^{\pm i\theta/3} = s_1, s_2$.
    c) $2(-p)^{1/2} \cos(\theta/3) = s_1 + s_2$.
    d) Use $e^{2\pi i/3} = \omega, e^{2\pi i/3} = \omega^2$ to finish.

11. (**) Derive the trigonometric solution from Cardan's solution. Hint: Define θ by $(-p)^{1/2} e^{i\theta/3}$ where $0 < \theta < \pi$, then show the following.
    a) $\cos\theta = -q/(-p)^{3/2}$.
    b) $s_2 = (-p)^{1/2} e^{i\theta/3}$.
    c) $r_1 = s_1 + s_2 = 2(-p)^{1/2} \cos(\theta/3)$.
    d) Use $\omega = e^{2\pi i/3}$, $\omega^2 = e^{2\pi i/3}$ to finish.

12. (***) The trigonometric solution is not (strictly speaking) an algebraic solution. Show that we cannot solve the general equation of degree $n > 4$ even if we allow the use of trigonometric functions along with the elementary algebraic operations.

7. Algebraic Solutions of Polynomials

7.4 Solution of the General Quartic

Let $P(x)$ be a quartic polynomial. Without loss of generality we may assume that $P(x)$ is monic and the coefficient of x^3 is zero, so

$$P(x) = x^4 + cx^2 + dx + e = 0.$$

We rewrite this equation as

$$(x^2)^2 = -cx^2 - dx - e. \tag{7.3}$$

If the right side of this expression was a perfect square, then we could take the square root and solve two quadratics for the zeros of $P(x)$. We work towards the right side being a square by adding the following to each side of Equation (7.3)

$$2\left(\frac{y}{2}\right)x^2 + \left(\frac{y}{2}\right)^2.$$

This leaves the left side of Equation (7.3) a square and allows us to choose y so that the right side is also a square. We will carry this out in a moment, but for now though let's just jump to the final result.

Theorem 7.4.1 (Ferrari's Solution to the General Quartic). *Let $P(x)$ be a monic quartic polynomial:*

$$P(x) = x^4 + bx^3 + cx^2 + dx + e.$$

*Let y be a zero of the **cubic resolvent** of $P(x)$:*

$$Q(y) = y^3 - cy^2 + (bd - 4e)y + (4ce - d^2 - b^2 e).$$

The zeros of $P(x)$ are the zeros of the following two quadratics:

$$2X^2 + \left(b + \sqrt{b^2 - 4c + 4y}\right)X + \left(y + \sqrt{y^2 - 4e}\right)$$

$$2X^2 + \left(b - \sqrt{b^2 - 4c + 4y}\right)X + \left(y - \sqrt{y^2 - 4e}\right).$$

where the square roots are chosen so that

$$\sqrt{b^2 - 4c + 4y}\sqrt{y^2 - 4e} = by - 2d.$$

The product of these two quadratics is $4P(x)$.

7.4. Solution of the General Quartic

If all the coefficients of P are real, then the computation may be simplified by choosing y to be the largest real zero of the resolvent. For this choice of y the coefficients of the quadratics will be all real.

Example 7.4.1. *Find the zeros of $P(x) = x^4 + 3x + 20$.*

Solution: For this polynomial $b = 0$, $c = 0$, $d = 3$, and $e = 20$, so the cubic resolvent is
$$y^3 - 80y - 9 = 0.$$
This has the zero $y = 9$ (isn't the author kind to himself?). The square roots
$$\sqrt{b^2 - 4c + 4y} = \pm 6 \quad \text{and} \quad \sqrt{y^2 - 4e} = \pm 1$$
must be chosen so that their product is $by - 2d = -6$. We choose the square roots to be $+6$ and -1 respectively. The quadratics are
$$2\mathbf{X}^2 + 6\mathbf{X} + 8 \quad \text{and} \quad 2\mathbf{X}^2 - 6\mathbf{X} + 10,$$
so have the zeros
$$\frac{3}{2} \pm \frac{\sqrt{11}}{2} i, \quad \text{and} \quad \frac{3}{2} \pm \frac{\sqrt{7}}{2} i.$$
These are the zeros of $P(x)$. ∎

Example 7.4.2. *Find the zeros of $P(x) = x^4 + x^3 + x^2 + x + 1$.*

Solution: For this polynomial $b = 1$, $c = 1$, $d = 1$, and $e = 1$, so the cubic resolvent is
$$y^3 - y^2 - 3y + 2 = 0.$$
This has the zero $y = 2$. The square roots
$$\sqrt{b^2 - 4c + 4y} = \pm\sqrt{5} \quad \text{and} \quad \sqrt{y^2 - 4e} = 0$$
clearly satisfy the requirement that $by - 2d = 0$. The quadratics are
$$2X^2 + (1+\sqrt{5})X + 2 \quad \text{and} \quad 2X^2 + (1-\sqrt{5})X + 2$$
which have the zeros
$$\frac{-1+\sqrt{5}}{4} \pm \frac{\sqrt{10+2\sqrt{5}}}{4} i, \quad \frac{-1-\sqrt{5}}{4} \pm \frac{\sqrt{10-2\sqrt{5}}}{4} i.$$
These are the zeros of $P(x)$. Compare this solution to that of Example 5.4.1 (page 81). ∎

7. Algebraic Solutions of Polynomials

Derivation of Ferrari's Solution

Again, we started by adjusting the quartic $P(x)$ so that it is monic and the coefficient of x^3 is zero,
$$P(x) = x^4 + cx^2 + dx + e = 0,$$
and then wrote this as
$$(x^2)^2 = -cx^2 - dx - e. \tag{7.4}$$

After adding
$$2\left(\frac{y}{2}\right)x^2 + \left(\frac{y}{2}\right)^2$$
to each side of Equation (7.4), we have
$$\left(x^2 + \left(\frac{y}{2}\right)\right)^2 = (y-c)x^2 - dx + \left(\frac{y^2}{4} - e\right). \tag{7.5}$$

A quadratic equation is a perfect square if and only if its discriminant is zero. The right side of Equation (7.5) then is a perfect square if and only if
$$(-d)^2 - 4(y-c)\left(\frac{y^2}{4} - e\right) = 0.$$

Rearranging, we find that y must satisfy the following cubic (which we called the resolvent of $P(x)$):
$$y^3 - cy^2 - 4ey + (4ec - d^2) = 0.$$

Let y be any zero of the resolvent, then the right side of Equation (7.5) is a square, say
$$(y-c)x^2 - dx + \left(\frac{y^2}{4} - e\right) = (\gamma x + \delta)^2 \tag{7.6}$$
for some choice of γ and δ. Comparing the coefficients of x^2 and 1, we see $\gamma = (y-c)^{1/2}$ and $\delta = \left(\frac{y^2}{4} - e\right)^{1/2}$. Finally, comparing the coefficients of x, we find
$$2\gamma\delta = 2(y-c)^{1/2}\left(\frac{y^2}{4} - e\right)^{1/2} = -d.$$

That is, the two square roots must be chosen so that their product is $-d$. Combining Equations (7.5) and (7.6), and taking the square root, we find that x must satisfy one of the following quadratics:
$$x^2 + \frac{y}{2} = \pm(\gamma x + \delta).$$

After multiplying by 2 and rearranging, these become the two quadratics in Ferrari's solution:
$$2x^2 \pm (4y - 4c)^{1/2}x + \left(y \pm (y^2 - 4e)^{1/2}\right) = 0.$$

Exercises 7.4:

1. Find the zeros of the following polynomials.

    a) $x^4 - 6x^3 + 12x^2 - 20x - 12$

    b) $x^4 + x^3 - x^2 - 7x - 6$

    c) $x^4 - 11x^2 - 6x + 10$

    d) $x^4 - 3x^2 - 12x + 40$

    e) $x^4 + 2x^3 - 7x^2 - 8x + 12$

    f) $x^4 + 11x^2 + 10x + 50$

    g) $x^4 + 2x^3 - 10x^2 - 8x + 24$

    h) $x^4 - 6x^3 + 12x^2 - 14x + 3$

    i) $x^4 - 2x^2 + 8x - 3$

    j) $x^4 - 8x^3 + 21x^2 - 26x + 14$

    Hint: To simplify the computation, most of the resolvents of these examples have rational zeros.

2. By direct multiplication (and using the resolvent), show that the product of the two quadratics in Ferrari's solution is $4P(x)$.

3. Prove that if the coefficients of $P(x)$ are real and we choose y to be the largest real zero of the resolvent of $P(x)$, then the coefficients of the two quadratics in Ferrari's solution are real.

4. Suppose that we are using Ferrari's solution and accidently choose the signs of the square roots incorrectly. Show that we find the negative of the zeros of $P(x)$ instead of the zeros of $P(x)$.

5. (*) The derivation we gave for Ferrari's solution assumes that b (the coefficient of x^3) is zero. Derive the solution without this assumption. Hint: Follow the the derivation above.

6. (*) Find the general solution to equations of the form
$$ax^8 + bx^7 + cx^6 + dx^5 + ex^4 + dx^3 + cx^2 + bx + a.$$

    Hint: Divide by x^4 and let $y = x + x^{-1}$. Find $x^m + x^{-m}$ in terms of y for $m = 2, 3$ and 4.

7. (*) Write a program to find the solution of the general quartic with real coefficients. Hint: Begin with the programs that solve the general (real) cubic and the general quadratic. With a minimal amount of additional programming these two programs will be sufficient. Notice y may always be chosen to be real.

Geometric Constructions

*It is easier to square the circle
than to get round a mathematician.*

Augustus de Morgan

*In most sciences one generation tears down what
another has built and what one has established
another undoes. In Mathematics alone each
generation builds a new story to the old structure.*

Herman Hankel

Chapter 8

Geometric Constructions

In this chapter we use the theory of equations to solve several construction problems from classical Greek geometry. For example, using just an unmarked ruler and compass, can we draw a regular polygon with seventeen sides? We all know how to bisect an angle, but can we trisect one (divide an arbitrary angle into three equal angles)? Here geometry, when mixed with complex numbers and algebra, together are capable of a great feat: determining exactly which constructions are, and which are not, possible. That is the story we tell in this chapter.

So after laying the historical foundation in Section 8.1, we start as usual in mathematics, with careful definitions. In Section 8.2 we define the geometric notion of constructability and of constructible points. We then translate this geometric concept of constructability into an analytic concept by showing how to associate the constructible points with certain complex numbers—called constructible numbers. In Section 8.4, we study the polynomials with constructible numbers as zeros, where we finally can give a rule for deciding which constructions are possible. In Section 8.5 we apply these ideas to the classical problems and present several generalizations. Finally we deviate from the ancient Greeks and discuss geometry with origami in the final section.

8.1 Squaring the Circle

THE STANDARD TOOLS of classical geometry are the straight edge and compass. Euclid in his great classic "The Elements" presented many of the standard constructions of geometry, showing for example how to construct an equilateral triangle, a square, a regular pentagon, a hexagon, and the regular polygon with fifteen sides. Euclid also presented the standard methods for bisecting an angle, constructing perpendicular lines, constructing line segments with given lengths, and even constructing a square with the same area as a given triangle, rectangle or polygon. What Euclid

8. Geometric Constructions

could not answer, and neither could any other mathematician for over two thousand years, is how (or even if) we can do the following:

Square the Circle Given a circle, construct a square with the same area.

Duplicate the Cube Given a cube, construct a cube with exactly twice the volume.

Trisect an Angle Given any angle, divide it into three equal sub-angles.

Construct the Regular n-gon Given any integer n greater than two, construct the regular polygon with n sides.

Constant searching for solutions over such a long period resulted in many fruitful discoveries in mathematics. For example the ellipse, parabola and hyperbola are said to have been discovered by Menaechums in an attempt to solve these problems. A great part of the Theory of Equations and much of the spirit of Group Theory is linked with these problems.

One of the first persons to attempt to solve the trisection problem was Hippas of Elis (500BC), who found a solution by breaking the rules—using something other than a straight edge and compass. A partial list of those who worked on these problems attests to the tremendous appeal of these deceptively easily stated problems: Archimedes, Nicomedes, Pappus, Leonardo da Vinci, Dürer, Descartes, Ceva, Pascal, Huygens, Leibniz, Newton, Maclaurin, Mascheroni, Gauss, Steiner, Chasles, Sylvester, Kempe, Klein, Dickson. The search was by no means left only in the hands of the great; for example in 1775 the Paris Academy passed a resolution that its members were to waste no more time examining supposed solutions to the problems of squaring the circle, duplicating the cube, trisecting an angle or building a perpetual motion machine! Some amateurs still continue these fruitless searches! See for example Underwood Dudley's fine text on trisectors [14].

The first three problems were shown to have no solution around 1800. About the same time, it was shown that a regular polygon of $n > 2$ sides is constructible if and only if

$$n = 2^j F_1 F_2 \ldots F_m$$

where j and m are non-negative integers, and the F_i are distinct Fermat primes. (The Fermat Primes are the primes of the form $2^{2^k} + 1$; as of 2015 the only known Fermat primes are 3, 5, 17, 257 and 65537 ($k = 0, 1, 2, 3$ and 4).)

Why did it take 2000 years to resolve these problems? To answer these questions, mathematicians had to reason outside the confines of plane geometry—introducing both analytic and algebraic concepts which did not exist when the questions were first asked.

It is interesting to note that now, more than two centuries after the first three problems have been shown to have no solution, supposed solutions are still being circulated! Most mathematics professors have met at least one person offering a solution to the trisection problem, often expecting instant acclaim, fame, and hopefully even money. These trisectors usually have had little or no mathematical training, so few can be convinced that their task is not possible, and indeed many seem to feel that it is not possible to prove any task impossible. Many trisectors feel that they are a last ray of hope and understanding in a world choked by lying mathematicians. For example, a trisector who happened to be the president of an American university wrote in 1931:

> ... This age has gone further in this respect than any other; it has extended its attacks to the utmost bounds of science. The mutineers [modern mathematicians] against the old order have seized the ship of knowledge and nailed the flag of dissent to the mast; they have driven the defenders of all manner of orthodoxy below decks and battened down the hatches over them, and have left in their administration not a single department of science. ... When normally sound criticism turns into destructive bolshevism, it is time to consider whether the criticism is as sound as that which it criticizes.

For more information on trisectors see *The Trisection Problem* by R. Yates [32]. For an interesting history of these construction problems see *Squaring the Circle* by E. Hobson [21], and *Famous Problems* by F. Klein [24].

In this chapter we resolve each of the four construction problems. In section two we carefully define the notion of constructability and the set of constructible points. In section three we translate the geometric concept of constructability into an analytic concept by showing how to associate constructible points with certain complex numbers—called constructible numbers. This allows us to add, subtract, multiply, divide and even take the square roots of constructible points. In section four we translate the notion of constructability into an algebraic concept by studying the polynomials with constructible numbers as zeros. We state a simple rule for deciding which constructions are possible based on the degrees of these polynomials and use this to show the first three constructions above are impossible. In section five we use this rule to decide which regular polygons are constructible.

(At least) **read the exercises in this chapter.** In sections 2, 3, 4 and 5 much of the material is developed in the exercises.

Exercises 8.1:

1. Look up Euclid's Postulates (Axioms). Which of them allow us to use a straight edge, and which allow the use of a compass?

2. Show there are only 31 (known) constructible regular polygons with an odd number of sides. How many would be known if we find a sixth Fermat prime?

8. Geometric Constructions

3. Using something in addition to a ruler and compass, give at least two ways of exactly trisecting the angle.

4. (*) Many people are still attempting to trisect the angle or carry out other constructions that have been shown impossible. Why? (Consider [14].)

5. (*) Write a short essay on the history of one of the four problems.

8.2 The Geometric Concept

The rules of classical (Greek) geometry are said to be handed down from Plato and allow only the use of a straight edge and compass. The **straight edge** (also called a **ruler**) is not a measuring device, and may be used only for drawing lines. To stress this it is often called an *unmarked straight edge*. Similarly the compass is not a measuring device, and may be used only to draw circles. In fact, the compass is traditionally assumed to collapse when lifted from the paper, and is therefore called a *collapsible compass*. Given any set S of points **the only allowed operations are:**

- drawing the straight line through any two given distinct points;
- drawing the circle with a given point as center and passing through a second given point.

That's it! No marking your ruler or using your compass to measure a distance is allowed. When performing these operations we cannot pick random points, not even a random point on a given line or circle. The only points that may be used for these operations are:

- the original set S of given points,
- the point of intersection of any two lines we have drawn,
- the point(s) of intersection of any line and any circle we have drawn, and
- the point(s) of intersection of any two circles that we have drawn.

The points determined by a *finite* number of such operations are called **the points constructible from the set S**. In the special case that we are to construct something on an empty plane (no points given), we may assume the existence of exactly two points. The infinite set of points constructible from these two are called the **constructible points**.

- A *line* is constructible if and only if it contains two distinct constructible points.

8.2. The Geometric Concept

- A *line segment* is constructible if and only if the endpoints are constructible.

- A *circle* is constructible if and only if the center is constructible and the circle passes through a constructible point.

- A *polygon* is constructible if and only its vertices are constructible points.

- An angle θ is constructible if and only if there are two constructible lines such that the angle between them is θ.

In general, any figure is **constructible by ruler and compass** if and only if (1) the figure consists only of points, line segments, lines and circles; and (2) each point, line segment, line and circle in the figure is constructible.

We close this section with three of the basic constructions of Euclidean geometry, constructions which we will use often in this chapter.

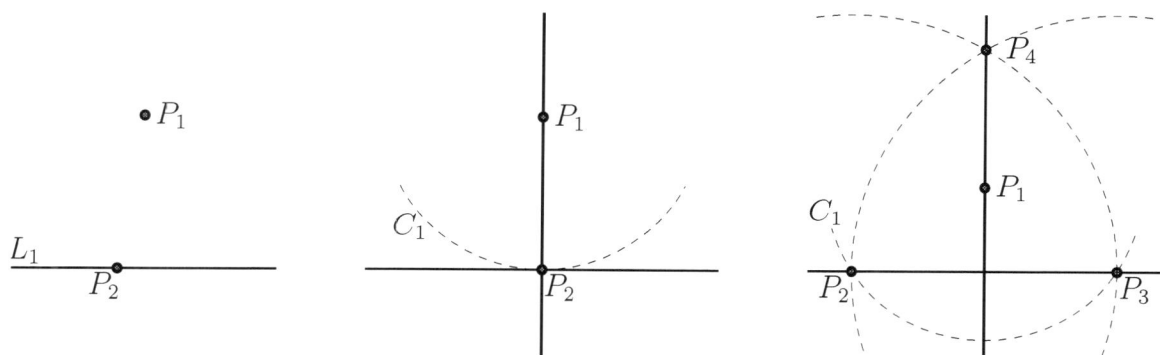

Figure 8.1: Constructing a Perpendicular Through a Given Point

Example 8.2.1. *Given a line L_1 and a point P_1, construct the line L_2 through P_1 perpendicular to L_1.*

Solution: The expression "given line L_1" means that we are given at least two (constructible) points on the line L_1. Let one of them be called P_2 (choose $P_1 \neq P_2$). See Figure 8.1, left.

Draw the circle C_1 with center P_1 passing through P_2. If the circle is tangent to L_1 then the desired line is the line through P_1 and P_2 (Figure 8.1, center).

If the circle is not tangent to the line, then the circle C_1 intersects L_1 in two distinct points, P_2 and P_3 (Figure 8.1, right). Draw two further circles: the circle with center P_2 passing through P_3, and the circle with center P_3 passing through P_2. These two circles intersect in two points, P_4 and P_5. The line through P_4 and P_5 is the desired line. ■

8. GEOMETRIC CONSTRUCTIONS

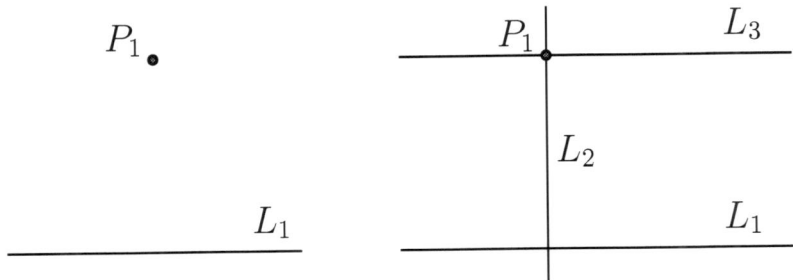

Figure 8.2: Constructing a Parallel Through a Given Point

Example 8.2.2. *Given a line L_1 and a point P_1 not on L_1, draw the line through P_1 parallel to L_1.*

Solution: We will just use the previous construction twice. Construct the line L_2 through P_1 perpendicular to L_1 (Figure 8.2). The line through P_1 perpendicular to L_2 is the desired line. ∎

Since our ruler is unmarked and our compass is collapsible, we cannot use them to "move" lengths; but our next example shows there is a way of measuring off a previously constructed length on a given line.

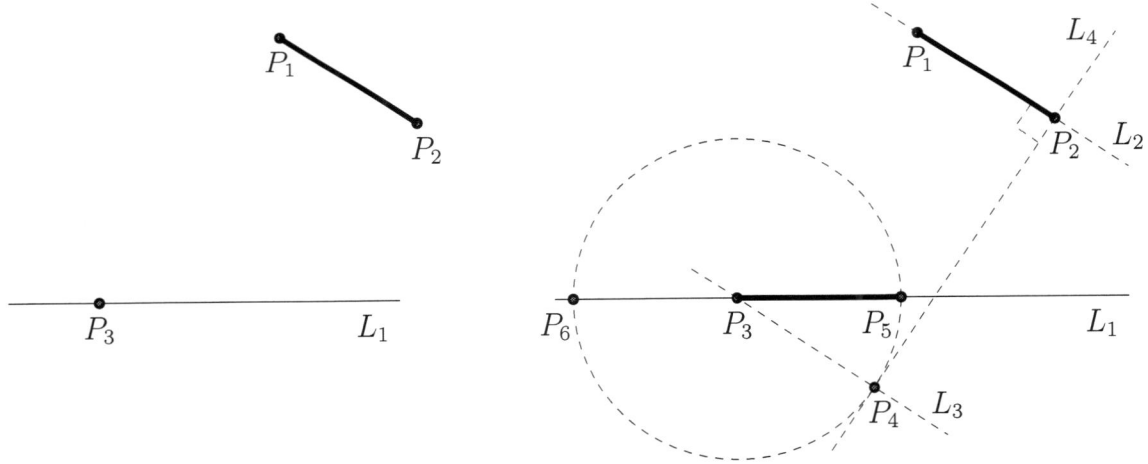

Figure 8.3: "Moving" a Line Segment

Example 8.2.3. *Given a line segment $\overline{P_1P_2}$, a line L_1, and a point P_3 on L_1, construct the two line segments on L_1 which have as one endpoint P_3, and have the same length as $\overline{P_1P_2}$. (That is, "move" the line segment $\overline{P_1P_2}$ onto the line L_1 with one endpoint on P_3. See Figure 8.3.)*

8.2. The Geometric Concept

Solution: The case that P_1, P_2, and P_3 are collinear is left to the reader (Exercise 9). Let the line through P_1 and P_2 be called L_2 (Figure 8.3). Draw the line L_3 through P_3 and parallel to L_2. Draw the line L_4 through P_2 perpendicular to L_2. The lines L_3 and L_4 are perpendicular, so they intersect; call the point of intersection P_4. Finally draw the circle with center P_3 passing through P_4. This circle intersects L_1 at two points, P_5 and P_6. So $\overline{P_3P_5}$ and $\overline{P_3P_6}$ are the desired line segments. ∎

Using GeoGebra

It is especially pleasant to use computer geometry systems such as GeoGebra to do these constructions. GeoGebra is free and has a large user support community. Particularly nice for teachers is that the menus can be simplified to (almost) just the standard Euclidean operations. For example, given two points you can draw the circle with one as center passing through the other, then when you move the initial points, the circle moves and changes size so that it is always centered on the one and passing through the other. So once a construction like Example 8.2.3 is done with GeoGebra, we can move the initial points and see how the construction moves and adapts. This ability to change the initial conditions after the construction is completed is a wonderful teaching tool and we recommend that many of the exercises in this chapter be done with GeoGebra. These exercises are marked with ▹.

Exercises 8.2:

1. ▹ Given a line segment $\overline{P_1P_2}$ show how to bisect it (that is, construct the midpoint P_3 of $\overline{P_1P_2}$).

2. ▹ Given a constructible angle show how to bisect it (that is, divide the angle into two equal sub-angles).

3. ▹ Given line segments with lengths x and y, show how to construct line segments with lengths:

    a) $x + y$ Hint: Example 8.2.3.

    b) $x - y$ given $x > y$.

    c) $1/x$ Hint: Similar triangles, Figure 8.4.

    d) $x \cdot y$ Hint: Similar triangles, Figure 8.5.

    e) x/y Hint: Similar triangles, Figure 8.6.

4. ▹ With a ruler and compass (starting with two points) construct the following:

    a) an equilateral triangle;

    b) a equilateral hexagon;

    c) a square;

8. GEOMETRIC CONSTRUCTIONS

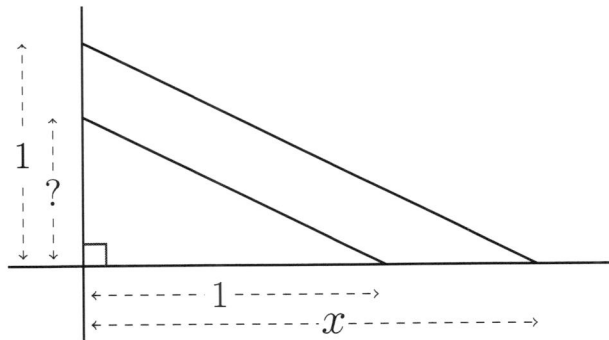

Figure 8.4: Construction Six (Problem 3)

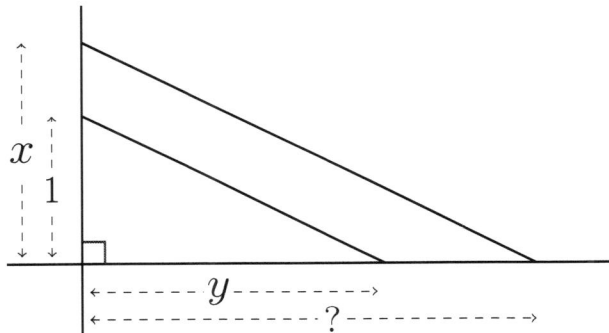

Figure 8.5: Construction Seven (Problem 3)

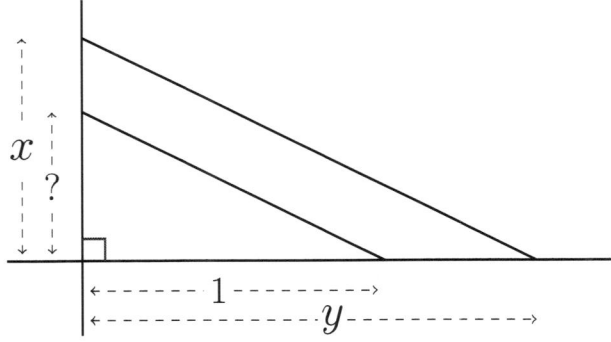

Figure 8.6: Construction Eight (Problem 3)

d) a regular octagon;

e) an isosceles triangle with sides equal twice the base;

f) a right triangle with sides 3, 4, 5; and

g) a triangle with sides 4, 5, 6.

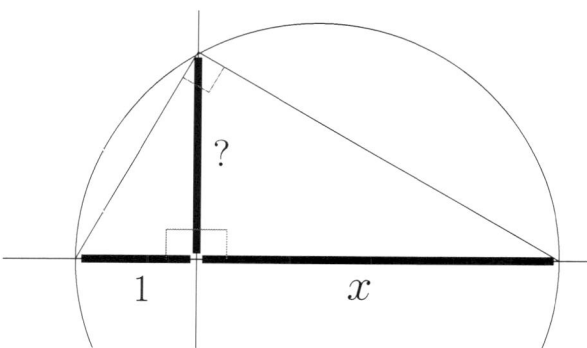

Figure 8.7: Constructing a Square Root (Problem 5)

5. (*) Given a line segment of length x, show how to construct one of length $x^{1/2}$. Hint: Figure 8.7.

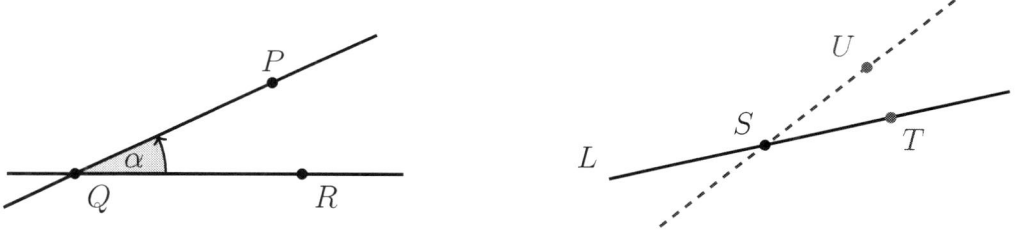

Figure 8.8: "Moving" an Angle α (Problem 6)

6. (*) Given an angle $\angle PQR$ defined by three points and a point S on a line L (Figure 8.8), find a point T on L and a point U so that $\angle UST \cong \angle PQR$.

7. Use the previous problem to show that if two angles are constructible, so is the angle that is the sum (or difference) of their measures.

8. Repeat the constructions of this section using GeoGebra: (a) 8.2.1 (b) 8.2.2 (c*) 8.2.3.

9. Show how to do the construction in Example 8.2.3 when P_1, P_2 and P_3 are collinear.

10. Given any triangle, construct (a) the circle through the vertices of the triangle, (b) the circle through the midpoints of the sides.

8. GEOMETRIC CONSTRUCTIONS

11. ⚐ Given three distinct points on a circle, show how to construct its center.

12. Show a circle is constructible if and only if it contains (goes through) three constructible points. Hint: The previous problem shows the if part.

13. Show that a circle C with center P and radius R is constructible if and only if P is constructible and there is a constructible line segment with length R.

14. Show that a polygon is constructible if and only if its sides are constructible line segments.

15. (*) Show that if we drop the restriction that our constructions are finite, then every point is constructible.

16. (*R) Prove that the set of constructible points is countable.

17. (A) A real number r is said to be constructible if and only if $|r|$ is the length of a constructible line segment (the unit for measurement is the distance between the initial two points). Show that it does not matter which line segment we pick as the unit of measurement.

18. (A) Show the set of lengths of constructible line segments is a multiplicative group (the unit for measurement is the distance between the initial two points). Hint: Exercise 3.

19. (*R) Show the set of constructible real numbers is a ring. Hint: Exercise 3.

20. In Example 8.2.1 we began with the points P_1 and P_2 (ignore any other given points). Show that the number of points constructed in this example is seven if P_1 is on L_1 and as follows if P_1 is not on L_1:

    one   if C_1 is tangent to L_1;
    nine   if C_1 is not tangent to L_1 and $P_1P_2P_3$ is an equilateral triangle; *or* if C_1 is not tangent to L_1, and $P_1P_2P_4$ is an isosceles triangle; and
    ten   otherwise.

8.3 The Analytic Concept

There is an obvious relationship between: the Euclidean plane on which we are doing our constructions, the complex plane, and the Cartesian plane—they are all planes! We may exploit this and associate the constructible points with a subset of the complex numbers, allowing us to perform algebraic operations on the set of constructible points. It is by adding this additional structure to the set of constructible points that we will be able to resolve the four classical problems.

8.3. The Analytic Concept

In determining which points are constructible points we began with exactly two points. Let's call one of these points **the origin** or **0**, and the other **1**. The distance between these points is our unit of distance. Call the line through 0 and 1 the **real axis** or **x-axis**, and call the line perpendicular to the real axis at 0, the **imaginary axis** or **y-axis**. See Figure 8.9.

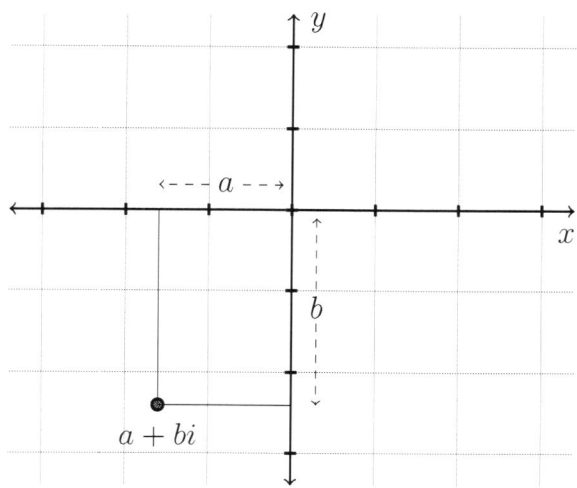

Figure 8.9: The Complex Plane $\mathbb{C}$

Now any point P corresponds to a unique complex number $a + bi$ and a unique ordered pair (a, b). A complex number $a + bi$ is a **constructible number** if and only if the point to which it corresponds is a constructible point. Notice that $a + bi$ is a constructible number if and only if $|a|$ and $|b|$ are the lengths of constructible line segments (exercise 1). Using this with exercises 3 and 5 of the previous section we may easily show the following.

Theorem 8.3.1. *If x and y are constructible numbers and $x \neq 0$, then the following numbers are also constructible:*

a) $Re(x)$ d) $1/x$ g) $x \cdot y$
b) $Im(x)$ e) $x + y$ h) y/x
c) $-x$ f) $x - y$ i) $\sqrt{x}$

Notice in particular that the integers and the rational numbers are constructible. Using Theorem 8.3.1 and the quadratic equation we can show the following (exercise 4).

147

8. Geometric Constructions

Corollary 8.3.2. *If a, b, and c are any constructible numbers, then the zeros of $ax^2 + bx + c = 0$ are also constructible.*

In short we have shown that we can add, subtract, multiply, divide and take the square roots of constructible numbers. In a strong sense the converse is also true.

Theorem 8.3.3 (The analytic concept of constructability). *A number z is constructible if and only if it can be expressed with a finite number of additions, subtractions, multiplications, divisions, and square roots beginning with the numbers one and zero.*

Notice that "beginning with the numbers one and zero" may be replace by "beginning with any two or more rational numbers" (exercise 5).

Proof. The 'if' condition is obvious from Theorem 8.3.1. To show the 'only if' we use our association of the complex numbers $x + yi$ with the ordered pairs (x, y). With this association we see that the line $y = ax + b$ is constructible if and only if a and b are constructible (exercise 6). Similarly the circle $(x - a)^2 + (y - b)^2 = r^2$ is constructible if and only if a, b and r are constructible (exercise 7). The constructible points are the points found by a finite number of intersections of lines and circles. That is, the coordinates may be found by solving a finite number of linear equations (intersection of two lines) and quadratic equations (intersections of a circle with a line or circle). Thus the coordinates of a constructible point can be found with a finite number of the algebraic operations above. □

Example 8.3.1. *Show that the equilateral triangle, the square, the pentagon and the hexagon are constructible.*

Solution: A regular polygon of n sides is constructible if and only if the angle $2\pi/n$ is constructible (exercise 8), which is constructible if and only if $\cos(2\pi/n)$ is constructible (exercise 9).

$\cos \frac{2\pi}{3} = -\frac{1}{2}$ so the equilateral triangle is constructible.

$\cos \frac{2\pi}{4} = 0$ so the square is constructible.

$\cos \frac{2\pi}{5} = \frac{\sqrt{5}-1}{4}$ so the pentagon is constructible.

$\cos \frac{2\pi}{6} = \frac{1}{2}$ so the hexagon is constructible.

We give explicit constructions of these polygons in Section 8.5. ■

8.3. The Analytic Concept

Exercises 8.3:

1. Prove that $a + bi$ is constructible if and only if there are constructible line segments of length $|a|, |b|$. Hint: For the 'if' part use example 8.2.3 of the previous section. For the 'only if' part draw the line perpendicular to the real axis through $a + bi$.

2. (*) Prove (a)–(h) of Theorem 8.3.1. Hint: (a)–(c) are the previous problem. For (d)–(h) first prove the results for real numbers using example 8.2.3 and exercise 3 of the previous section. Then prove the result for complex numbers by using (d)–(h) (on real numbers) and relations similar to the following:

$$\frac{a+bi}{c+di} = \frac{ac+bd}{c^2+d^2} + \frac{bc-ad}{c^2+d^2}i$$

3. (*) Prove (i) of Theorem 8.3.1. Hint: Let $(c+di)^2 = a+bi$, show that $4d^4 + 4ad^2 - b^2$ is zero, so $d^2 = (-a + \sqrt{a^2+b^2})/2$ where d is real. Use the previous problem; Exercise 5 and Example 8.2.3 of the previous section to show that d is constructible. Then $c = b/2d$ is constructible.

4. Use Theorem 8.3.1 to prove Corollary 8.3.2.

5. (*) Show that in Theorem 8.3.3 "beginning with the numbers one and zero" may be replaced by "beginning with any two or more rational numbers." Hint: Prove the rational numbers are constructible from 0 and 1 by using Theorem 8.3.1. Why is this enough?

6. Prove that a line $y = ax + b$ is constructible if and only if a and b are constructible. Hint: Show that this is equivalent to the line going through two constructible points.

7. Prove that a circle $(x-a)^2 + (y-b)^2 = r^2$ is constructible if and only if a, b and r are constructible. Hint: Use problem 13 of the previous section.

8. Show that a regular polygon of n sides is constructible if and only if the angle $2\pi/n$ is constructible. Hint: Let P_0 be the center, and P_1, P_2 two adjacent vertices of the polygon. What is the angle between the lines segments $\overline{P_0P_1}$ and $\overline{P_0P_2}$?

9. Show that an angle θ is constructible if and only if $\cos\theta$ is constructible. Hint: Draw a unit circle centered on the vertex of the angle.

10. (*) Show that the trigonometric function cosine may be replace by any of the other five trigonometric functions in the previous problem.

11. Show that a constructible circle can be 'squared' if and only if $\sqrt{\pi}$ is constructible.

12. Show that a constructible cube can be 'duplicated' if and only if $\sqrt[3]{2}$ is constructible.

8. Geometric Constructions

8.4 The Algebraic Concept

In the last section we showed that a point $z = a + bi$ is constructible if and only if z has the form

$$a + b_1\sqrt{c_{1,1} \pm \sqrt{c_{1,2} \pm \sqrt{\cdots}}} + b_2\sqrt{c_{2,1} \pm \sqrt{c_{2,2} \pm \sqrt{\cdots}}} + \cdots$$

where

$$a, b_1, b_2, \ldots, c_{1,1}, c_{1,2}, \ldots, c_{2,1}, \ldots$$

is a *finite* list of rational numbers (Theorem 8.3.3). We may now show the following.

> **Theorem 8.4.1 (The algebraic concept of constructability).** *A complex number z is constructible if and only if it is the zero of an irreducible polynomial P with rational coefficients and degree a power of two.*

It is beyond the scope this book to prove the sufficiency of the condition in this theorem. The proof belongs to a branch of Mathematics known as Galois Theory, and is covered in advanced texts on Abstract Algebra (Field Theory). For the necessity of the condition see exercise 16. It is important to note that in this section we use only the necessity of the condition.

The idea behind the proof of necessity may be illustrated by the following example.

Example 8.4.1. *The number $\sqrt{2} + \sqrt{3} + \sqrt{5}$ is constructible. Find an polynomial with this number as a zero.*

Solution: Set $x = \sqrt{2} + \sqrt{3} + \sqrt{5}$ and work to remove the radicals. To do this, we first remove the square root of 5 by subtracting $\sqrt{2} + \sqrt{3}$ from each side of the equality (isolating $\sqrt{5}$) and squaring:

$$(x - \sqrt{2} - \sqrt{3})^2 = x^2 - 2(\sqrt{2} + \sqrt{3})x + (5 + 2\sqrt{6}) = 5.$$

Notice that since we have squared this expression, it would not matter if the square root of five was replaced by its negative, that is, $x_1 = \sqrt{2} + \sqrt{3} - \sqrt{5}$ will also be a zero of the polynomial which we are finding. We now remove the square roots of three which occur in the coefficients by isolating these (coefficients in which a square root of three occurs) on the right side:

$$x^2 - 2\sqrt{2}x = \sqrt{3}\left(2x - 2\sqrt{2}\right),$$

and squaring,

$$x^4 - 4\sqrt{2}x^3 + 8x^2 = 12x^2 - 24\sqrt{2}x + 24.$$

8.4. The Algebraic Concept

Again we have squared so the square root of three could be replaced by its negative, that is, $x_2, x_3 = \sqrt{2} - \sqrt{3} \pm \sqrt{5}$ will also be zeros of the polynomial which we are finding.

We finally remove the square roots of two by isolating these on the right hand side $x^4 - 4x - 24 = \sqrt{2}(4x^3 - 24x)$ and squaring to get

$$x^8 - 8x^6 - 32x^4 + 192x^2 + 576 = 2\left(16x^6 - 192x^4 + 576x^2\right).$$

So the desired polynomial is

$$x^8 - 40x^6 + 352x^4 - 960x^2 + 576.$$

Once more, since we have squared, the square root of two may be replaced by its negative and the eight zeros of the polynomial above are as follows:

$$\sqrt{2} + \sqrt{3} + \sqrt{5}, \quad \sqrt{2} + \sqrt{3} - \sqrt{5}, \quad \sqrt{2} - \sqrt{3} + \sqrt{5}, \quad \sqrt{2} - \sqrt{3} - \sqrt{5},$$
$$-\sqrt{2} + \sqrt{3} + \sqrt{5}, \quad -\sqrt{2} + \sqrt{3} - \sqrt{5}, \quad -\sqrt{2} - \sqrt{3} + \sqrt{5}, \quad -\sqrt{2} - \sqrt{3} - \sqrt{5}.$$

Note that we could also have found the polynomial directly from this list of zeros. ∎

Notice the two key elements of the previous example: first, the polynomial was formed by rearranging and repeated squaring; second, because of these squarings each square root in the expression for x may be replaced by its negative, doubling the number of zeros. These are the key facts in proving the theorem. For a complete proof see the references [24], [21] or [25].

Recall that a polynomial with rational coefficients is **reducible** if it is the product of two other polynomials with rational coefficients and lower degree (that is, if it can be factored.) If a polynomial is not reducible then it is **irreducible**. Theorem 8.3.1 may be used to show the following (exercise 17).

Corollary 8.4.2. *If the number z is constructible and z is the zero of the irreducible polynomial $P(x)$, then $P(x)$ has degree a power of two.*

Example 8.4.2. *Show that it is not possible to duplicate a constructible cube.*

Solution: Let the side of the given cube be L. The volume of the cube is L^3. The side of the cube with twice the volume is $\sqrt[3]{2}\,L$. So to duplicate the cube we must be able to construct $\sqrt[3]{2} = \sqrt[3]{2}\,L/L$ (Theorem 8.3.1, last section). Notice $\sqrt[3]{2}$ satisfies the polynomial $x^3 - 2 = 0$, which is irreducible by exercise 1. This polynomial has degree 3 (not a power of 2), thus duplicating the cube is not possible. ∎

Example 8.4.3. *Show that it is not possible to trisect the general angle with ruler and compass.*

8. GEOMETRIC CONSTRUCTIONS

Solution: We show that the trisection is not possible in general by showing an angle of 60 degrees cannot be trisected. We can do this by showing an angle of 20 degrees cannot be constructed, or equivalently that $x = 2\cos 20°$ is not a constructible number (problem 9, page 149). Recall from trigonometry that

$$\cos 3\theta = 4\cos^3\theta - 3\cos\theta.$$

Multiply this identity by two and rearrange the terms to find

$$(2\cos\theta)^3 - 3(2\cos\theta) - 2\cos 3\theta = 0.$$

Setting $\theta = 20$ degrees, and $x = 2\cos\theta$ this identity becomes

$$x^3 - 3x - 1 = 0.$$

This polynomial is irreducible by problem 1, hence x is not constructible and general trisection is not possible. ∎

Recall that a number is **transcendental** if it not the zero of any polynomial with integral coefficients. For example, e, π and $\sqrt{\pi}$ are transcendental. We will discuss transcendental numbers and prove these numbers are transcendental in Chapter 11. An immediate consequence of the theorem above is as follows.

Corollary 8.4.3. *Transcendental numbers are not constructible.*

Example 8.4.4. *Show that it is not possible to square a constructible circle.*

Solution: The area of a circle is πr^2. The side of a square with the same area is $\sqrt{\pi}\, r$. If the given circle is constructible then r is constructible—so to construct the square we must be able to construct $\sqrt{\pi}$. But $\sqrt{\pi}$ is transcendental and cannot be constructed. ∎

Exercises 8.4:

1. Prove the following theorem and apply it to $x^3 - 2$ and $x^3 - 3x + 1$.

Theorem 8.4.4. *Let $Q(x)$ be a cubic polynomial with rational coefficients. Either $Q(x)$ has a rational zero or $Q(x)$ is irreducible.*

Hint: If $Q(x)$ is reducible it must be be the product of a linear polynomial and a quadratic polynomial both with rational coefficients (why?) hence has a rational zero.

8.4. The Algebraic Concept

2. Prove Corollary 8.4.3.

3. Find a polynomial with integer coefficients and

    a) the zero $2^{1/2} - 3^{1/2}$,

    b) the four zeros $\pm 2^{1/2} \pm 3^{1/2}$,

    c) the eight zeros $\pm 2^{1/2} \pm 3^{1/2} \pm 7^{1/2}$.

4. (*) Find the monic integral quartic with the zero $a^{1/2} + b^{1/2}$ (where $a \neq \pm b$ are integer variables).

5. (*) Show that the polynomial that we found in example 8.4.1 is irreducible.

6. The polynomial $x^8 - 44x^6 + 438x^4 - 1292x^2 + 529$ has the eight zeros $\pm\sqrt{2} \pm \sqrt{3} \pm \sqrt{6}$. However this polynomial is not irreducible; it factors as

$$(x^4 - 22x^2 - 48x - 23)(x^4 - 22x^2 + 48x - 23).$$

    Which of the eight zeros are the zeros of the first of these two factors? What pattern do you see in the signs of the square roots?

7. The polynomial $x^8 + 28x^6 + 294x^4 - 932x^2 + 2401$ has the eight zeros $\pm\sqrt{2} \pm \sqrt{3}\,i \pm \sqrt{6}\,i$ and is reducible. Find the factor with the zero $\sqrt{2} + \sqrt{3}\,i + \sqrt{6}\,i$.

8. Show that it is not possible to rectify a constructible circle. (To **rectify a circle** means to construct a line segment with length equal to the circumference of the circle.) Hint: Use the fact that π is transcendental.

9. We have seen that angles of 60 and 72 degrees are constructible (indirectly via Example 8.3.1), but an angle of 20 degrees is not. Use these to show that an angle of 3 degrees is constructible—but angles of 1 and 2 degrees are not.

10. Use the previous exercise to prove the following.

Theorem 8.4.5 (Constructible Angles). *Let n be an integer. The angle n degrees is constructible if and only if 3 divides n.*

11. Use the previous exercise to prove the following.

Corollary 8.4.6. *Let n be an integer. The angle n degrees is trisectable if and only if 9 divides n.*

12. Show that given any triangle, it is possible to construct a square with the same area. That is, it possible to square a triangle. Hint: How do you find the area of a triangle? Use Theorem 8.3.1.

8. GEOMETRIC CONSTRUCTIONS

13. (*) Using GeoGebra and starting with three points (the vertices of a triangle), construct a square with the same area as the triangle.

14. (*) Given the origin, the axes, and one point P, use GeoGebra to construct a square with the same area as the rectangle with vertices P and the origin, with sides parallel to the axes.

15. (**) Given any (finite) polygon bounded by constructible lines, show that it is possible to construct a square with the same area. This result helps explain the interest in squaring the circle. Hint: Show the following result. Let A be any positive real number. Then it is possible to construct a square with area A if and only if A is a constructible number.

16. (***) Prove that the condition of the Theorem 8.4.1 is necessary. Hint: Every constructible number x has the form

$$a + b_1\sqrt{c_{1,1} \pm \sqrt{c_{1,2} \pm \sqrt{\cdots}}} + b_2\sqrt{c_{2,1} \pm \sqrt{c_{2,2} \pm \sqrt{\cdots}}} + \cdots$$

with a, b_i, and $c_{i,j}$ rational. Define the **rank** n of the number x to be the greatest number of square roots in any term of x. So x equals

$$a + b_1 B_1^{1/2} + b_2 B_2^{1/2} + \ldots + b_n B_n^{1/2}$$

where $B_1, \ldots, B_n$ are numbers with rank at most $n-1$, and this equation may be viewed as a polynomial in x with coefficients of rank at most n. Using the technique of example 8.4.1, show how to remove the square roots and get a polynomial in x with degree a power of two and with coefficients of rank at most $n-1$. Use induction. What are the other zeros of the resulting polynomial?

17. (*) Prove Corollary 8.4.2.

8.5 Regular Polygons

Theorem 8.5.1. *A regular polygon of $n \geq 3$ sides is constructible if and only if*

$$n = 2^j F_1 F_2 \ldots F_m$$

where j, m are non-negative integers and the F_i are distinct Fermat primes.

This theorem was first proved on March 30th, 1796 by Carl Friedreich Gauss (1778-1855). The discovery of this result induced the 19-year-old Gauss to study mathematics instead of philology. The discovery so intrigued him, that he asked that it

8.5. Regular Polygons

be inscribed on his tombstone. We leave the proof of this theorem to the reader (exercises 1 through 7).

Recall a **Fermat number** is a number of the form

$$F_n = 2^{2^n} + 1.$$

A Fermat number that is prime is called a **Fermat Prime**. Fermat (1601-1665) was convinced that all Fermat numbers were prime, though he only knew that the first five such numbers were prime:

$$F_0 = 2^{2^0} = 3, \quad F_1 = 2^{2^1} = 5, \quad F_2 = 2^{2^2} = 17, \quad F_3 = 2^{2^3} = 257, \quad F_4 = 2^{2^4} = 65537.$$

These are still the only known Fermat primes; the next two dozen Fermat numbers have been shown to be composite. In particular, $F_5 = 4294967297$ was shown to have the factor 641 by Euler in 1732. In 1880, Landry discovered how to factor F_6 by hand!

$$F_6 = 274177 \cdot 67280421310721.$$

In 1971, Brillhart used a computer to discover that

$$F_7 = 59649589127497217 \cdot 5704689200685129054721.$$

Even if we cannot factor a given Fermat number, there is a relatively easy test to see if F_n is prime or not.

Theorem 8.5.2 (Pepin's Test). *F_n is prime if and only if F_n divides $3^{(F_n-1)/2} - 1$.*

(For a proof see [27, theorem 55].) It is possible to carry out this computation with numbers roughly the size of F_n^2, but due to the very large size of these numbers, very few F_n have been tested.

Currently F_n is known to be composite for n from 8 through 32—the smallest Fermat number F_n for which the nature is still undecided is F_{33}, a number with 2,585,827,973 digits! Many believe that the number of Fermat primes is finite (see for example [19, Section 2.5]). If we have not found them all, the next one will allow a n-gon with so many sides that we would not be able to tell it from a circle on this page with an electron microscope! For example, look at the 257-gon in figure 8.10, this is really a figure with 257 sides, not a circle. Yet even when we magnified one a small portion by a factor of 20, it is still difficult to see the line segments that made it.

Example 8.5.1. *Find all known odd integers n such that the regular n-gon is known to be constructible.*

8. Geometric Constructions

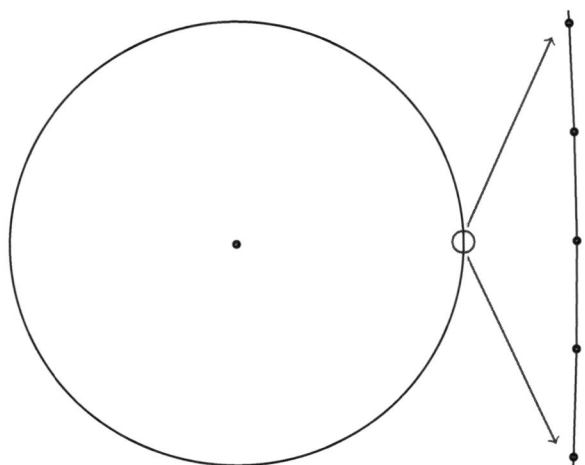

Figure 8.10: The 257-gon

Solution: By Theorem 8.5.1 n is a product of distinct Fermat primes. Since there are only 5 known Fermat primes $(F_0, \ldots, F_4)$ we know n is one of

$$\begin{array}{ccccc}
F_0, & F_1, & F_2, & F_3, & F_4, \\
F_0F_1, & F_0F_2, & F_0F_3, & F_0F_4, & F_1F_2, \\
F_1F_3, & F_1F_4, & F_2F_3, & F_2F_4, & F_3F_4, \\
F_0F_1F_2, & F_0F_1F_3, & F_0F_1F_4, & F_0F_2F_3, & F_0F_2F_4, \\
F_0F_3F_4, & F_1F_2F_3, & F_1F_2F_4, & F_1F_3F_4, & F_2F_3F_4, \\
F_0F_1F_2F_3, & F_0F_1F_2F_4, & F_0F_2F_3F_4, & F_1F_2F_3F_4, & F_0F_1F_2F_3F_4.
\end{array}$$

That is, n is one of the following numbers.

3	5	15	17
51	85	255	257
771	1285	3855	4369
13107	21845	65535	65537
196611	327685	983055	1114129
3342387	5570645	16711935	16843009
50529027	84215045	252645135	286331153
858993459	1431655765	4294967295	

■

It is interesting to note that Theorem 8.5.1 proves that it was possible to construct polygons of 3, 5, 17, 257 and 65537 sides—but gives us no hint how to perform these constructions. The constructions of the 3-gon (equilateral triangle) and the 5-gon (pentagon) may both be found in Euclid's Elements, but before Gauss' discovery, who would have even attempted a 17, 257 or 65537-gon?

8.5. Regular Polygons

Example 8.5.2. *Construct the regular pentagon.*

Solution: Beginning with the assumed two points A, B we construct a circle with center A and perpendicular radii $\overline{AB}$, $\overline{AC}$ (Figure 8.11). Bisect $\overline{AC}$ to find D. Draw the circle with center D passing through B to find E. Draw the circle with center B passing through E. The intersections of this circle and the point B are three of the vertices of the pentagon.

To prove that this is a regular pentagon, we first note that the right triangle DAB has sides $\frac{1}{2}$, 1, and $\frac{\sqrt{5}}{2}$. The right triangle EAB has sides 1, $\frac{\sqrt{5}-1}{2}$, and $\frac{5-\sqrt{5}}{2}$. From either example 5.4.1 or 7.4.2, we see the vertex P_1 of a regular pentagon is at

$$\frac{-1+\sqrt{5}}{4} + \frac{\sqrt{10+2\sqrt{5}}}{4} i.$$

A simple check shows the distance from this point to 1 is exactly the distance from B to E, so the construction above finds P_0, P_1 and P_4 exactly. Finally a circle with center P_1 (or P_4) passing through P_0 will intersect our original circle at P_2 (or P_3 respectively). ■

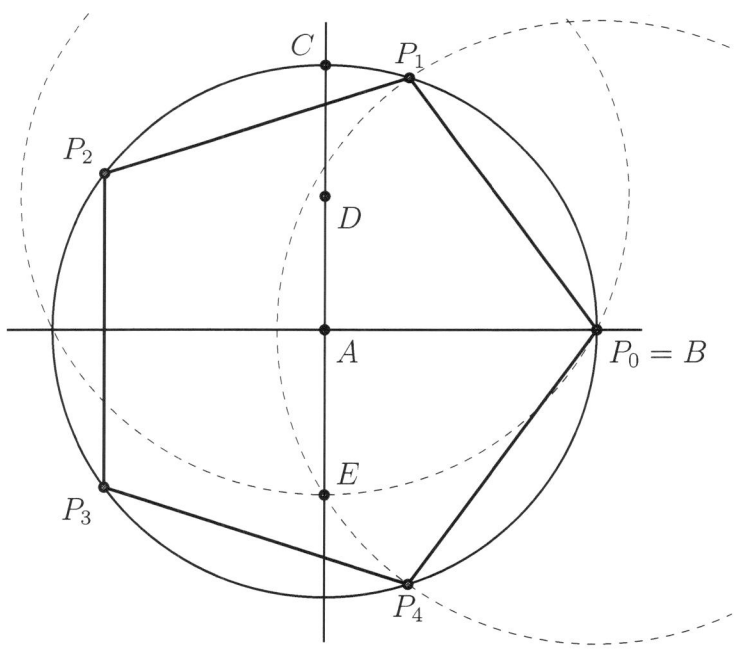

Figure 8.11: Constructing a Pentagon

8. Geometric Constructions

Example 8.5.3. *Construct the regular hexagon.*

Solution: Beginning with the assumed two points A, B we construct a circle with center A and passing through B. Let C be the second intersection of this circle with the line $\overleftrightarrow{AB}$ (Figure 8.12). Draw the circle with center C and passing through A, voilá. The proof of correctness is trivial. ∎

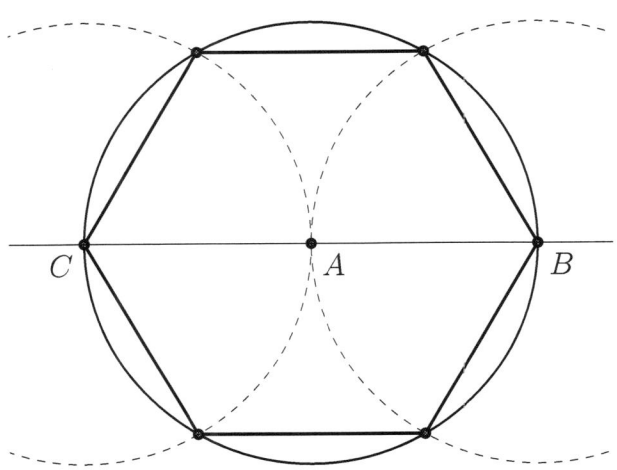

Figure 8.12: Constructing a Hexagon

Example 8.5.4. *Construct the regular 20-gon and the regular 30-gon.*

Solution: To construct the 20-gon we first construct the pentagon. Let P_0 and P_1 be any two consecutive vertices of the pentagon and A the center of the circle through its vertices (Figure 8.13). Bisect the side $\overline{P_0P_1}$, finding the point Q. Draw the line through A and Q, finding the intersection P_2 with the circle. $\overline{P_0P_2}$ is the side of the 10-gon. We then bisect the line segment $\overline{P_0P_2}$ to find Q', extend the diameter of the circle through Q' to find P. $\overline{P_0P}$ is the side of the 20-gon (the other sides are easily found). Notice the technique used here will work in general—given two vertices P_1, P_2 of an m-gon, bisect this side to find the $2m$-gon, then the $4m$-gon, ..., then the $2^n m$-gon.

To construct the 30-gon we first construct in the unit circle the pentagon (hence a 72 degree angle) and the hexagon (hence a 60 degree angle), each with a vertex at $1 = P_0$. Let P_0C, P_0D be a side of the pentagon and hexagon respectively. Then $\overline{CD}$ is a side of the 30-gon (see Figure 8.13); in fact, $C = P_6$ and $D = P_5$. The other sides are now easily found.

A similar method may be used to construct the mn-gon given the m-gon and n-gon (when m, n are relatively prime). ∎

8.5. Regular Polygons

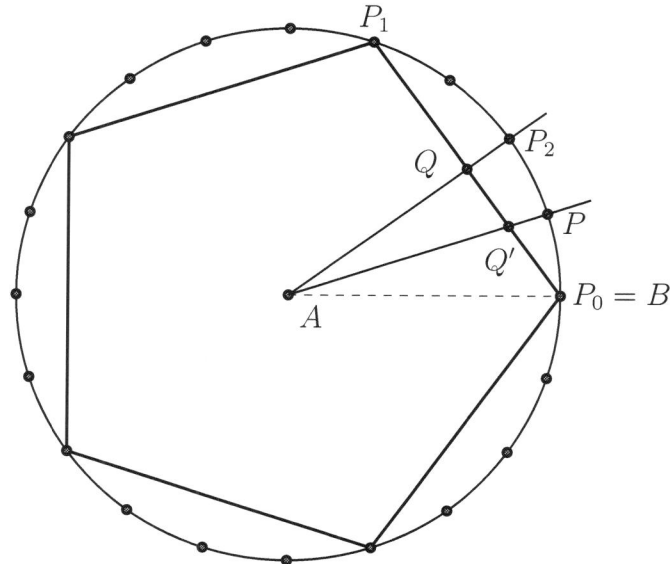

Figure 8.13: Constructing a 20-gon

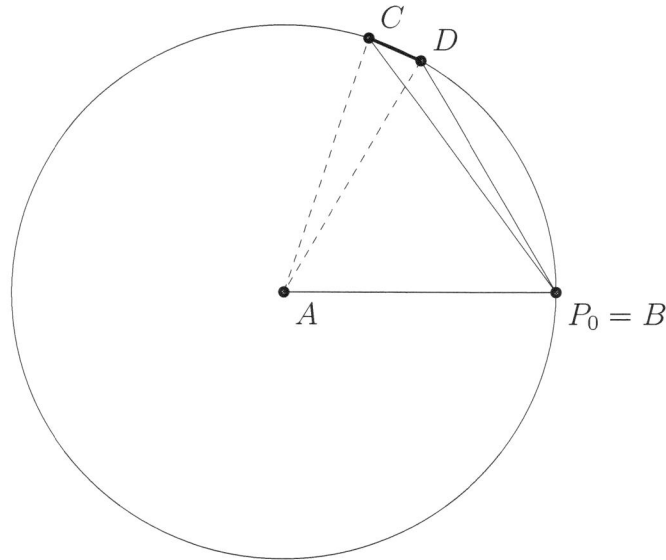

Figure 8.14: Constructing a 30-gon

8. Geometric Constructions

Example 8.5.5. *Construct the regular 17-gon.*

Solution: (This is a construction due to Richmond.) Start with a circle with center A and the two perpendicular radii AB and AC. Bisect AC twice to find $AD = (1/4)AC$ (see Figure 8.15). Bisect the angle ADB twice finding the angle ADE. Draw a perpendicular at D to DE, bisecting the right angle to find H such that the angle EDH is 45 degrees. Draw the circle with diameter BH to find its intersection K with AC. Draw the circle with center E passing through K to find N_3 and N_5 on the line $\overleftrightarrow{AB}$. The lines perpendicular to $\overleftrightarrow{AB}$ at N_3, N_5 intersect the original circle at the 3rd, 14th and 5th, 12th vertices of the 17-gon respectively. The rest are easily found. ∎

For the construction of the 257-gon (or 65537-gon) see *How to Construct a Regular Polygon* by W. Bishop [9].

Exercises 8.5:

1. (The first problems lead the reader through a proof of Theorem 8.5.1.) Show that the regular n-gon is constructible if and only if one of the following equivalent conditions hold.

    a) The angle $\dfrac{2\pi}{n}$ is a constructible angle.

    b) The regular n-gon may be inscribed in the unit circle with one vertex at 1.

    c) The nth zeros of unity are constructible numbers.

2. (*) Let m, n be relatively prime integers. Show that if both the regular n-gon and m-gon are constructible, then so is the regular nm-gon. Hint: Since n, m are relatively prime, by the extended Euclidean Algorithm, there exist integers a, b such that $an + bm = 1$. Notice $\dfrac{2\pi}{n}b + \dfrac{2\pi}{m}a = \dfrac{2\pi}{nm}$.

3. Show that if the regular n-gon is constructible, then so is the regular $2n$-gon. Conclude that for any positive integer k, the regular $2^k n$-gon is constructible. Hint: Bisect.

4. Show that if n and m are positive integers such that the regular nm-gon is constructible, then the regular n-gon and m-gon are constructible. Hint: Every n^{th} vertex of the regular nm-gon forms a _____.

5. Use the previous three exercises to prove the following.

Lemma 8.5.3. *Let $n = 2^k F_1^{m_1} \cdots F_n^{m_n}$, where the F_i are distinct odd primes. The regular n-gon is constructible if and only if the regular $F_i^{m_i}$-gon is constructible for each i.*

8.5. Regular Polygons

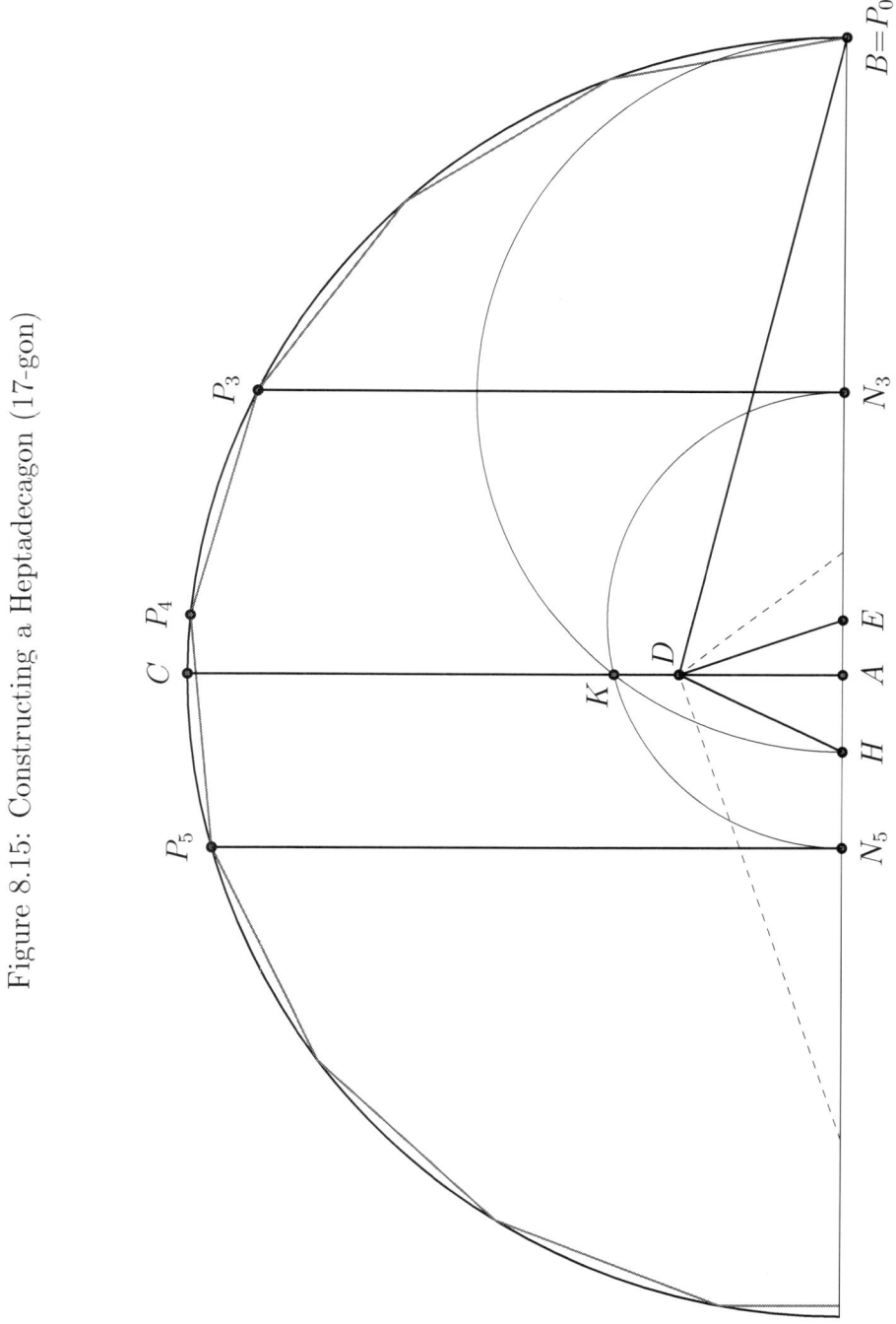

Figure 8.15: Constructing a Heptadecagon (17-gon)

8. GEOMETRIC CONSTRUCTIONS

6. Show that if F is an odd prime, then the regular F^2-gon is not constructible. Hint: The F^2 zeros of unity are the zeros of the irreducible polynomial (Example 5.3.3):

$$\frac{x^{F^2}-1}{x^F-1} = x^{F(F-1)} + x^{F(F-2)} + \ldots + x^F + 1.$$

Use Corollary 8.4.2.

7. Show that if F is an odd prime, then the regular F-gon is constructible if and only if $F = 2^m + 1$ for some integer m. Hint: The Fth zeros of unity are the zeros of the irreducible polynomial:

$$(x^F - 1)/(x - 1) = x^{F-1} + x^{F-2} + \ldots + x + 1.$$

Use Corollary 8.4.2.

8. Show that if $F = 2^m + 1$ is prime, then $m = 2^r$ for some integer r. Hint: if m has an odd factor u (say $m = uv$), then show that F may be factored because

$$2^m + 1 = (2^v + 1)(2^{m-v} - 2^{m-2v} + 2^{m-3v} - \ldots - 2^v + 1).$$

Use the previous four problems to prove Theorem 8.5.1.

9. ▷ Carry out the constructions for the following regular n-gons: (a) the pentagon; (b) the hexagon; (c) the 10-gon; (d) the 12-gon; (e) the 15-gon; (f*) the 20-gon; (g*) the 30-gon; (h**) the 17-gon; (i**) the 34-gon.

10. Show that the number of digits in (the base ten representation of) F_n is the greatest integer less than or equal to $1 + 2^n \log 2$.

11. (*) One of the largest Fermat numbers known to be composite is $F_{2478782}$. Let M be the number of digits in $F_{2478782}$. How many digits are there in M?

12. Let N be the next (integer) year in which an N-gon can be constructed. Show that $N = 2040$.

13. (*) Show that
$$F_0 F_1 F_2 \ldots F_n + 2 = F_{n+1}$$

Hint: Induction.

14. (*) Show that if there are m Fermat primes, then there are $2^m - 1$ constructible regular n-gons with n an odd integer.

15. Find all constructible regular n-gons (n odd or even) with less than (a) 10; (b) 100; (c) 1000; and (d*) 10^{30} sides.

8.6 Closing Comments

In this chapter we started with four ancient problems. Rather than concentrate on these four problems, we considered ruler and compass constructions in general and began by carefully defining constructible points. We then added structure to this set by showing how to associate them with complex numbers—then with the zeros of certain polynomials allowing a complete solution. It is very common in mathematics to start with definitions, experiment with generalization, then add extra structure; but this chapter presented one of the most powerful and sweeping examples of this process available at the undergraduate level! Remember that in mathematics the method of solution is often far more important and enlightening than the solution itself, and this is the case with this chapter.

Geometric Prices!

A favorite problem of mine starts by fixing a price to drawing a circle and to drawing a line, then asking you to complete a standard construction as cheaply as possible. For example, if a circle cost ten cents and a line five cents, how could you draw a hexagon for less than seventy cents? As a high-school student I always enjoyed competing at these type of competitions with my fellow students because these problems allow the students to use both their geometry knowledge and their own ingenuity. Changing the prices dramatically changes the construction and allows students to research the best possible constructions for each pricing scheme.

As an extreme example, suppose circles were free. Then every construction would be free because Mascheroni and Adler have shown any points constructible by ruler and compass are constructible by compass alone! Of course without a ruler we cannot draw a line, but we can construct every point on that line which is constructible. If instead lines were free, then every construction could be done for the cost of one circle because any constructible point can also be constructed by ruler alone if we are first given a single fixed circle. Both of these results may be found in fine book *Squaring the Circle* [21].

Bending the Rules

If we bend (or break) the rules, then almost every construction is possible. A surprising amount of constructions become possible if we mark our rulers, or use a carpenter's square or a compass that does not collapse. We can create devices to trisect angles like the "tomahawk" from Figure 8.16 (see problem 10). We present just a couple of these in the exercises below.

Archimedes gave a particularly simple trisection of an angle using a ruler with two marks. Place one mark of the ruler at P to find a point Q on one side of the angle, then draw the circle with radius $\overline{PQ}$ and center Q. This crosses the other ray

8. Geometric Constructions

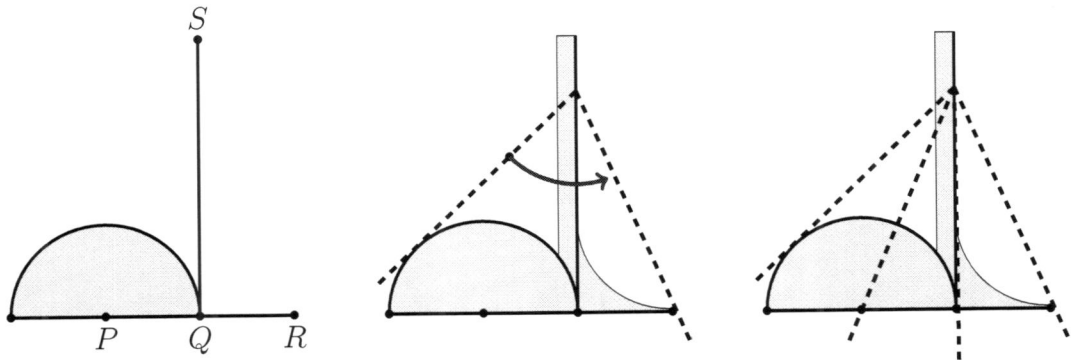

Figure 8.16: Using a Tomahawk to Trisect an Angle (Problem 10)

of the angle at a point R (see Figure 8.17). Draw the line through Q parallel to $\overline{PR}$. Now use the ruler to draw the line through P for which the one mark of the ruler lies on the circle and the other on the parallel line. Call the point of intersection of the two lines T. Now $\angle QPR = 3\angle TPR$. (The simple proof is left to the reader as an problem 7.)

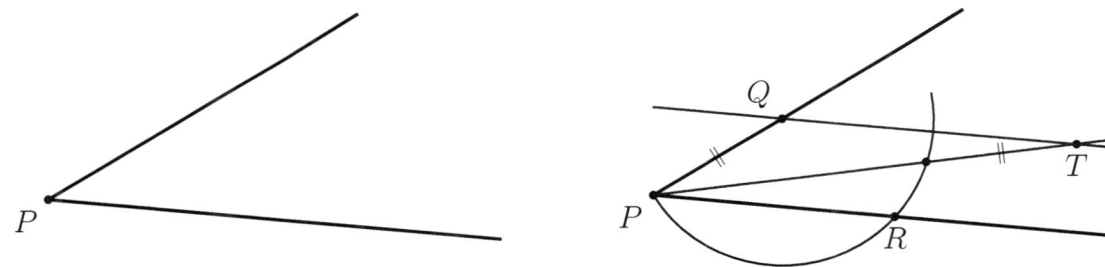

Figure 8.17: Archimedes Trisection with a Marked Ruler (Problem 7)

Below we focus on a particularly pleasant way of bending the rules: origami.

Using Origami

In Section 8.2 we pointed out that classical geometry only allows two basic operations: given two distinct points either drawing the line through them or drawing a circle with one as center through the other. Using origami, we cannot draw circles, but we can fold lines, and because of the flexibility of the paper, this gives a wide range of basic operations:

1. given two points P and Q, there is a unique fold that passes through both of them;

2. given two points P and Q, there is a unique fold that places P onto Q;

3. given two lines L_1 and L_2, there is a fold that places L_1 onto L_2;

4. given a point P and a line L, there is a unique fold perpendicular to L that passes through point P;

5. given two points P and Q and a line L_1, there is a fold that places P onto L_1 and passes through Q;

6. given two points P and Q and two lines L_1 and L_2, there is a fold that places P onto L_1 and Q onto L_2 simultaneously;

7. given one point P and two lines L_1 and L_2, there is a fold that places P onto L_1 and is perpendicular to L_2.

Get a piece of paper and try doing each of the above. These are often called the axioms of origami geometry.

The key difference between classical constructions and origami is the second to last operation. This move makes it possible to trisect angles, to calculate cube roots, and to solve cubic equations. If you want to more know about using origami to teach mathematics (not just geometry!), see Thomas Hull's excellent text "Project Origami' [22]. Hull's text presents trisecting the angle (exercise 11) and constructing $\sqrt[3]{2}$ (exercise 13) as classroom activities with student worksheets and even pedagogical explanations for the teachers. Hull [23] even showed it is possible to solve the general cubic using an origami version of Lill's method from Section 5.5!

Exercises 8.6:

1. Suppose you are charged 10 cents to draw a circle and 5 cents to draw a line. Show that it is possible to construct

    a) an equilateral triangle for less than 50 cents;

    b) a square for less than 90 cents;

    c) (*) a regular pentagon for less than $1.20;

    d) a regular hexagon for less than $1.00;

    e) (*) a regular 10-gon for less than $2.25.

    Try to get near (or beat) the possible prices 0.35, 0.60, 1.05, 0.65 and 1.50 dollars respectively. You also might also try varying the cost of a circle and line to see how the constructions must be altered to get the lowest price.

2. With compass alone construct (the vertices of): (a) an equilateral triangle; (b*) a square; (c*) a pentagon; (d) a hexagon. As usual we will start with two points and

you may add any lines necessary to illustrate your answer—but do not use them to find any intersections! Hint: For the pentagon the construction of the previous section need only be slightly modified.

3. Using the rules of the last problem and assuming the distance between the two initial points is one, explain how to construct (a) an angle of 30°; (b) the length $\sqrt{3}$; and (c*) the length $\sqrt{5}$.

4. Give one, two or three points on a plane, show that it is not possible to construct any further points with ruler alone.

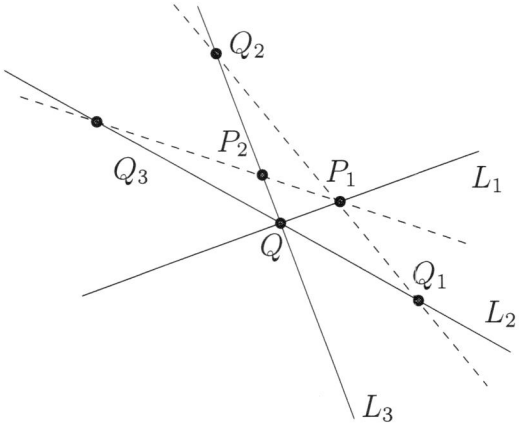

Figure 8.18: Constructing Lines Only (Problem 5)

5. (*) Prove the following.

Lemma 8.6.1. *Given three lines L_1, L_2, L_3 through the point Q, and two further points on L_1, say Q_1 and Q_3, on opposite sides of Q, then it is possible to construct an infinite number of additional points using the ruler alone.*

Hint: Let Q_2 be another of the given points on L_3 (see Figure 8.18). Repeat the following: Let P_1 be the intersection of L_2 and the line segment Q_1Q_2. Let P_2 be the intersection of L_3 and Q_3P_1. Show P_2 is between Q_2 and Q. Repeat using P_2 instead of Q_2.

6. Use the previous problem to show that given four points, using ruler alone, we can construct either 0, 1 or an infinite number of additional points. Hint: 0 if any three points are collinear; 1 if the four points form a rectangle; infinity otherwise.

7. Prove that Archimedes trisection in Figure 8.17 exactly trisects the original angle. Hint: See Figure 8.19. When are angles $\angle TQS$, $\angle QTS$ and $\angle QPS$ equal? Since the sum of the angles in a triangle is π, what is the angle $\angle PSQ$?

8.6. Closing Comments

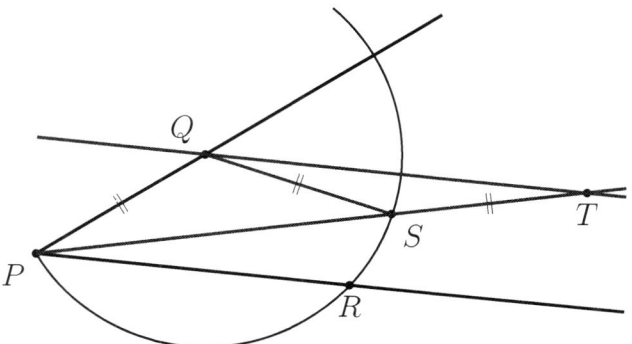

Figure 8.19: Archimedes Trisection Proof (Problem 7)

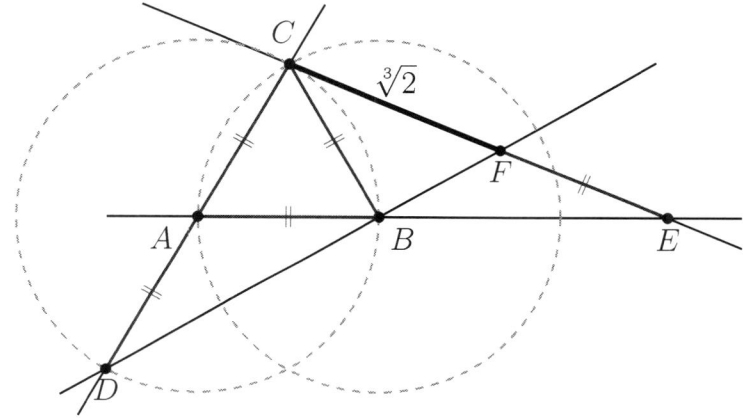

Figure 8.20: Constructing $\sqrt[3]{2}$ with a Marked Ruler (Problem 8)

8. (*) Show that the following construction with a marked ruler exactly finds $\sqrt[3]{2}$ (see 8.20). Start with an equilateral triangle $\triangle ABC$ whose side is the distance between the marks on the ruler. Use the marked ruler to extend the side $\overline{CA}$ a distance equal to the distance between the marks, finding D. Draw lines extending $\overleftrightarrow{DB}$ and $\overleftrightarrow{AB}$ as shown. Finally use the marked ruler to find the line through C intersecting $\overleftrightarrow{DB}$ at F and $\overleftrightarrow{AB}$ at E where the length of $\overline{FE}$ is the distance between the two marks. Now $\overline{CF}$ is $\sqrt[3]{2}$ times the distance between the marks on the ruler.

9. (*) Carpenters use an L-shaped "square" to make sure corners are built at right angles (see Figure 8.21). Given a circle, an unmarked carpenters square, and a pencil, explain how to construct the length of the side of a pentagon inscribed in the given circle and then prove your construction is correct. (Use Figure 8.21 in your explanation.)

Hint: First, explain how to find the point A and draw the two tangents to the circle. Now how from those line segments (the tangents) can you draw the two perpendicular diameters to find the points B and C? Finally, explain the use of the square with points

8. Geometric Constructions

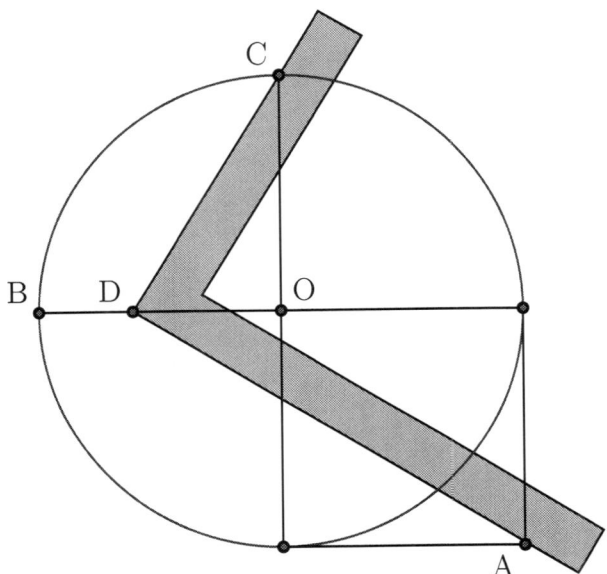

Figure 8.21: Carpenter's Square (Problem 9)

A and C to find D. It would be a good idea to make yourself a square and do this construction manually–then write up the steps. You do not need any marks on the square. One approach to showing the construction correct is to identify the point C with the point $(0,1)$, A with $(1,-1)$ and D with $(x,0)$. Find x, then the length of CD.

10. We can trisect angles by using non-Euclidean tools. A **tomahawk trisector** (see Figure 8.16) is formed using a semicircle with a diameter through $\overline{PQ}$ extended so that the length of $\overline{QR}$ equals the radius. Then a perpendicular to $\overline{PQ}$ is drawn at Q and we usually add some extra area to make something we can cut out of the paper and use (center image). To use the tomahawk, we place the vertex of the angle on $\overline{QS}$ and align it so one ray is tangent to the semicircle and the other ray passes through R (center image). Then the rays through P and through Q trisect the angle. Construct a tomahawk and prove that it trisects angles. Hint: Look for similar triangles.

11. Show that the origami method in Figure 8.22 trisects the given angle. Hint: Draw a vertical line through P and call the point that intersects the base of the paper S. Show the lengths of the segments $\overline{RQ}$, $\overline{QP}$ and $\overline{PS}$ are equal. Similar triangles .... (This method is due to H. Abe in 1980[7].)

12. The previous problem assumes that we can make two folds parallel to an edge which divides a paper into thirds. Show that the origami method in Figure 8.23 forms one of these folds. Hint: Let the lower left corner of the square be the point (0,0) and let the side of the square have length one. Now write down the equations of the lines from which we found R.

8.6. Closing Comments

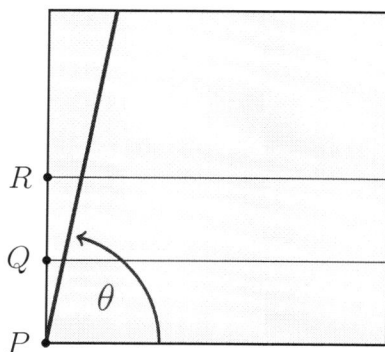 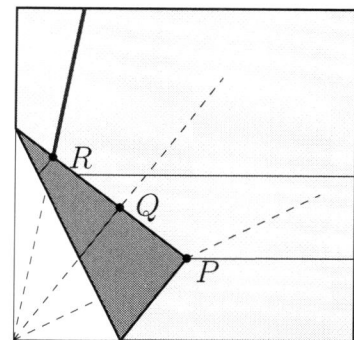

Form two horizontal folds by folding in half twice. Now fold R onto the line defining the angle and simultaneously place P onto the lower horizontal line. The fold line through Q extends trisecting the angle θ.

Figure 8.22: Trisecting an Angle with Origami (Problem 11)

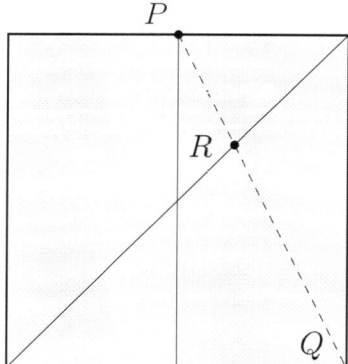 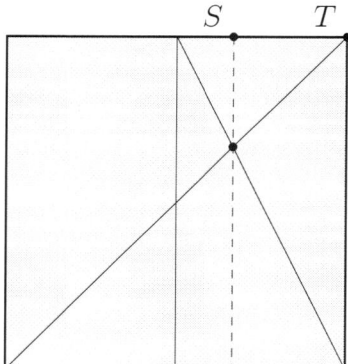

Fold the paper in half vertically and diagonally, then along the line segment $\overline{PQ}$ to find the intersection R. Now fold the right edge over (aligning the top edges) to form a vertical line through R. The point S where this line crosses the top is one-third of the length of the paper from T.

Figure 8.23: Trisecting a Length with Origami (Problem 12)

8. GEOMETRIC CONSTRUCTIONS

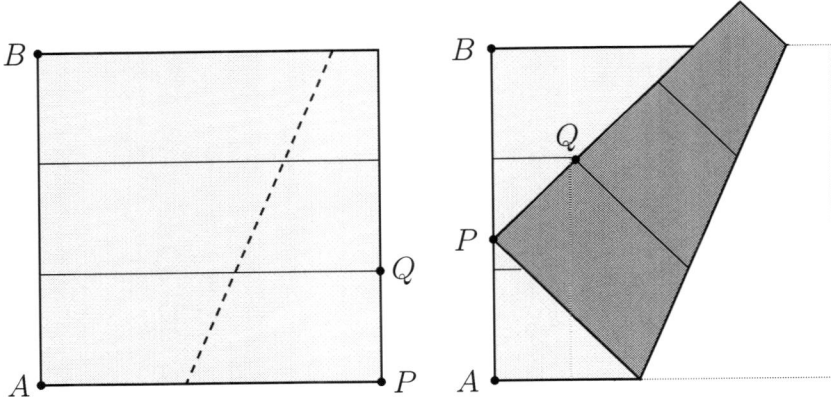

Fold the paper into three equal parts. Now fold the paper so that the point P lands on the left edge and the point Q falls on the line one-third of the paper from the top, then we have $\overline{PB} = \sqrt[3]{2}\,\overline{AP}$.

Figure 8.24: Constructing $\sqrt[3]{2}$ with Origami (Problem 13)

13. (*) The key to "duplicating the cube" is constructing $\sqrt[3]{2}$ which is not possible with ruler and compass alone. Show that the origami method in Figure 8.24 constructs the ratio $\sqrt[3]{2}$. Hint: It is rather difficult to do this geometrically, so introduce coordinates and try doing it algebraically.

14. A Pierpont prime is a prime of the form $2^n 3^m + 1$ for non-negative integers n and m. Those under 1,000 are: 2, 3, 5, 7, 13, 17, 19, 37, 73, 97, 109, 163, 193, 257, 433, 487, 577, and 769. (It is conjectured that there are infinitely many such primes.) For origami (or ruler, compass and angle trisector), Theorem 8.5.1 becomes the following.

Theorem 8.6.2. *A regular polygon of $n \geq 3$ sides is constructible by origami if $n = 2^j 3^k P$ where j, k are non-negative integers and P is a product of zero or more Pierpont primes.*

Find all of the origami-constructible regular polygons with at most 100 sides.

15. (*) Show that if we pick a line at random, the probability of it being constructible is zero. Hint: A constructible line has an algebraic slope—why? Use the following fact: If a number is picked at random, the probability that it is algebraic is zero.

16. (*) Show that if we pick a circle at random, then the probability that it is constructible is zero. Hint: Using the fact in the previous problem, prove the stronger result that a point picked at random has probability zero of being constructible. A constructible circle has a constructible center.

17. Write a concise outline of this chapter's solutions of the four classical problems mentioned at the beginning of this chapter.

Separating the Zeros
of Polynomials

The science of Pure Mathematics, in its modern developments, may claim to be the most original creation of the human spirit.

A. N. Whitehead

All the effects of nature are only the mathematical consequences of a small number of immutable laws.

P. S. Laplace

Mathematics—the unshaken Foundation of Sciences and the plentiful Fountain of Advantage to human affairs.

Isaac Barrow

Chapter 9

Separating the Real Zeros of Polynomials

If we are finding the zeros of a polynomial numerically, we need to know roughly where they are. Similarly, when we draw a graph of a polynomial, we want to show the "interesting parts" of the graph, and this means knowing roughly where its zeros (and perhaps the zeros of its derivatives) are. So in this chapter we present techniques to isolate, bound, and estimate the number of real zeros of polynomials. These vary from the quick and imprecise (such as Descartes' Rule of Sign), to the slow but sure (Sturm's Theorem).

9.1 Separating the Zeros

IN THE FOLLOWING CHAPTER we will "solve" polynomial equations numerically; that is, we will approximate their real zeros as opposed to exactly (or symbolically) determining them. To do this it helps to first **separate** (that is, isolate) the real zeros. We have separated a zero of a polynomial when we have found an interval (a, b) which contains that one zero of the polynomial, but no other zeros. We say the zeros are separated when each real zero is separated into its own interval. For example, the polynomial

$$4x^3 - 8x^2 - x + 2$$

has a zero in each of the intervals $(-1, 0)$, $(0, 1)$ and $(1, 3)$, so its zeros are separated.

To separate the (real) zeros of a polynomial it is important to know how many real zeros there are, roughly what size they are, and how we can determine how many are in any given interval. This chapter offers answers to each of these questions.

Let's start with a simple way to show that a polynomial has a zero in a given interval. Recall that polynomials are continuous functions, so their graphs have no

9. Separating the Real Zeros of Polynomials

gaps or breaks. One consequence of this is the following version of the Intermediate Value Theorem.

Theorem 9.1.1. *If $P(x)$ changes sign on an interval (a, b) (i.e., if $P(a)P(b) < 0$), then $P(x)$ has a zero in (a, b).*

Example 9.1.1. *Show that $P(x) = x^3 - 6x + 2$ has a zero in each of the intervals $(-3, 0)$, $(0, 1)$ and $(1, 3)$.*

Solution: We quickly find $P(-3) = -7$, $P(0) = 2$, $P(1) = -3$, and $P(3) = 11$. This shows $P(x)$ changes sign on each of the three given intervals. (Since this is a cubic, there are no other zeros.) ∎

Example 9.1.2. *Show that if $P(x)$ has odd degree, then $P(x)$ has at least one real zero.*

Solution: What we must show is that $P(x)$ has a zero in the interval from minus infinity to infinity. As we increase the size of x, the leading term $a_0 x^n$ of $P(x)$ dominates over the other terms. So for large x the sign of $P(x)$ is determined solely by the leading term. (In many algebra classes this is call the "endpoint behavior.")

Now consider the finite interval $(-a, a)$. If a is large enough, then $P(a)P(-a)$ has the same sign as $a_0 a^n a_0 (-a)^n = (-1)^n (a_0 a^n)^2$—which is negative since n is odd. So $P(a)$ and $P(-a)$ have opposite signs, showing $P(x)$ has a zero in $(-a, a)$ for some large value of a. ∎

Example 9.1.3. *Let $P(x)$ have even degree. Show that if the leading coefficient of $P(x)$ and its constant term have opposite signs, then $P(x)$ has at least two real zeros, one positive and one negative.*

Solution: Using the reasoning in example above, we know that for some large a both $P(-a)$ and $P(a)$ have the same sign as $a_0(\pm a)^n$. But n is even, so this sign is the same as the sign of the leading coefficient. We are also given that $P(0)$ (the constant term) has the opposite sign, so $P(x)$ has a zero in each of $(-a, 0)$ and $(0, a)$.
∎

We have now seen one way to show a polynomial has a zero in a given interval, but how do we know which intervals to check? One method is trial and error. We know the sign of the polynomial for large values from the leading term. We also always check the value when $x = 0$ because it is trivial to find.

9.1. Separating the Zeros

Example 9.1.4. *Separate the zeros of*

$$P(x) = 8x^4 + 16x^3 - 22x^2 - 24x + 15.$$

Solution: Note that $P(x)$ is positive when the absolute value of x is very large (so we say $P(x)$ is positive at $\pm\infty$). Note $P(0)$ is also positive, so we need to find points at which $P(x)$ is negative.

Next we try $x = \pm 1$ because these are also easy to evaluate. We see $P(1) < 0$, showing we have zeros in $(0,1)$ and $(1,\infty)$. Sadly $P(-1) > 0$, giving no additional information. Next we try $x = \pm 2$. Since $P(2) > 0$ we know that our second interval with a zero can be shortened to $(1,2)$. Finally, $P(-2) < 0$, so we also know that $P(x)$ has a zero in both $(-2,-1)$ and $(-\infty,-2)$.

We now have four intervals, each containing a zero. However $(-\infty,-2)$ is infinitely long. Noticing $P(-3) > 0$ allows us to shorten $(-\infty,-2)$ to $(-3,-2)$. We can summarize the information about the signs as follows.

value of x	$-\infty$	-3	-2	-1	0	1	2	3	∞
sign of $P(x)$	$+$	$+$	$-$	$+$	$+$	$-$	$+$	$+$	$+$

We have now shown the zeros of $P(x)$ are separated into the intervals

$$(-3,-2), \quad (-2,-1), \quad (0,1) \quad \text{and} \quad (1,2).$$

∎

There are many ways of separating the zeros without using trial and error. One way can be derived from the following theorem first proved by Rolle in 1691. (Some suggest this was known to Indian mathematician Bhāskara II before 1185!)

Theorem 9.1.2 (Rolle's Mean Value Theorem). *Between every two consecutive real zeros of $P(x)$ there is at least one zero of $P'(x)$.*

Example 9.1.5. *Separate the zeros of*

$$P(x) = 3x^4 - 4x^3 - 24x^2 + 48x - 17.$$

Solution: Differentiating we find

$$P'(x) = 12x^3 - 12x^2 - 48x + 48 = 12(x^2 - 4)(x - 1),$$

so $P'(x)$ has the zeros $1, 2, -2$. Constructing a table as in the previous example, we get the following.

177

9. SEPARATING THE REAL ZEROS OF POLYNOMIALS

value of x	$-\infty$	-2	1	2	∞
sign of $P(x)$	$+$	$-$	$+$	$-$	$+$

So the zeros of $P(x)$ are in

$$(-\infty, -2), \quad (-2, 1), \quad (1, 2), \quad \text{and} \quad (2, \infty).$$

■

The obvious problem with using Rolle's Theorem is that we separate the zeros of one polynomial by solving another. If $P'(x) = 0$ is also too difficult to solve directly, we have gained nothing.

Exercises 9.1:

1. Separate the zeros of the following polynomials. If there are no real zeros, then you are already done!

   a) $11x^4 - 31x^2 + 7$

   b) $x^3 - 6x + 2$

   c) $x^4 + 4$

   d) $x^4 - 10x^2 + 16$

   e) $x^4 - 2x^3 - 22x^2 + 40x + 40$

   f) $x^3 - x^2 - 9x + 8$

2. Prove Theorem 9.1.1.

3. Prove Theorem 9.1.2. Hint: Use the Mean Value Theorem from your calculus course.

> **Theorem 9.1.3 (Mean Value Theorem).** *If the function $f(x)$ is continuous on the closed interval $[a, b]$, and differentiable on the open interval (a, b), then there is a number c in (a, b) for which*
>
> $$f'(c) = \frac{f(b) - f(a)}{b - a}.$$

4. Use Theorem 9.1.2 to find conditions on p and q so that

$$x^3 - px + q = 0$$

has three real zeros. When done, compare your answer to discriminant of the cubic.

9.2 Descartes' Rule of Signs

In a sequence of numbers such as

$$a_0, \quad a_1, \quad a_2, \quad \ldots, \quad a_n,$$

none of which is zero, any two consecutive terms a_i, a_{i+1} will either have the same sign or opposite signs. If they have opposite signs we say they present a **change of sign**. If the sequence contains zero terms, then they are ignored. For example, in the sequence

$$27, \quad 9, \quad 0, \quad 2, \quad -23, \quad 23, \quad 0, \quad 0, \quad -3, \quad 0, \quad -3, \quad 3, \quad 2$$

there are four changes of sign. René Descartes gave the following theorem in his 1637 text *La Geometrie*.

Theorem 9.2.1 (Descartes' Rule of Signs). *The number of positive real zeros (counting their multiplicity) of the polynomial with real coefficients*

$$a_0 x^n + a_1 x^{n-1} + \ldots + a_n$$

is less than or equal to the number of changes of sign in its coefficients; and if the number of zeros is less, then it is less by an even number.

We delay the proof of this theorem until the appendix A.3. Note that Descartes' Rule counts the multiplicity of a zero. For example, the polynomial

$$(x-2)^4 = x^4 - 12x^3 + 96x^2 - 256x + 256$$

has only one real zero ($x = 2$) witch has multiplicity four. The coefficients of this polynomial present four changes of sign which matches this multiplicity.

Example 9.2.1. *Bound the number of positive and the number of negative zeros of the polynomial*

$$P(x) = x^5 + 2x^3 - x^2 + x - 1.$$

Solution: The coefficients of $P(x)$ present three changes of sign so $P(x)$ has either three or one positive zeros. The negative zeros of $P(x)$ are positive zeros of

$$P(-x) = -x^5 - 2x^3 - x^2 - x - 1.$$

Since $P(-x)$ has no change of sign, $P(x)$ has no negative zeros. ■

9. Separating the Real Zeros of Polynomials

With a bit of magic we can determine exactly the number of positive zeros of $P(x)$ in this last example using Descartes' Rule. Form the new polynomial

$$(x+1)P(x) = x^6 + x^5 + 2x^4 + x^3 - 1.$$

Multiplying by $x+1$ does not change the number of positive zeros because it only adds a new negative one. The resulting polynomial has only one change of sign, so both $(x+1)P(x)$ and $P(x)$ have exactly one positive, hence four imaginary, zeros. This is magic because finding a term like $x+1$ by which to multiply takes trial and error and in most cases does not even exist. Still, we thought it might be useful for you to see the trick at least once.

Isaac Newton proposed a rule for providing a lower bound on the number of imaginary zeros.

Theorem 9.2.2 (Newton's Incomplete Rule, 1681). *Let*

$$P(x) = a_0 \binom{n}{0} x^n + a_1 \binom{n}{1} x^{n-1} + a_2 \binom{n}{2} x^{n-2} + \ldots + \binom{n}{n} a_n$$

be a polynomial with real zeros and real coefficients. Then $P(x)$ has at least as many (non-real) complex zeros as there are variations in sign in the sequence:

$$a_0^2, \quad a_1^2 - a_2 a_0, \quad a_2^2 - a_3 a_1, \quad a_3^2 - a_4 a_2, \quad \ldots, \quad a_{n-1}^2 - a_n a_{n-2}, \quad a_n^2.$$

(For a proof see D.J. Acosta [2].)

Example 9.2.2. *Bound the number of positive and the number of negative zeros of the polynomial*

$$P(x) = 3x^5 + 2x^4 + 3x^3 + 2x^2 + 3x + 2.$$

Solution: By Descartes' Rule, we know this has no positive zero; and either 5, 3 or 1 negative zeros. To use Newton's Incomplete Rule, we write the polynomial as

$$P(x) = 3(1)x^5 + \tfrac{2}{5}(5)x^4 + \tfrac{3}{10}(10)x^3 + \tfrac{2}{10}(10)x^2 + \tfrac{3}{5}(5)x + 2(1).$$

The sequence of terms in Newton's Incomplete Rule are

$$3^2, \quad \left(\tfrac{2}{5}\right)^2 - (3)\left(\tfrac{3}{10}\right), \quad \left(\tfrac{3}{10}\right)^2 - \left(\tfrac{2}{5}\right)\left(\tfrac{2}{10}\right),$$

$$\left(\tfrac{2}{10}\right)^2 - \left(\tfrac{3}{10}\right)\left(\tfrac{3}{5}\right), \quad \left(\tfrac{3}{5}\right)^2 - \left(\tfrac{2}{10}\right)(2), \quad 2^2$$

which has signs

$$+, \quad -, \quad +, \quad -, \quad -, \quad +$$

9.2. Descartes' Rule of Signs

so this polynomial has at least four complex zeros. Together these show $P(x)$ has one negative and four complex zeros. In fact the zeros are exactly as follows.

$$-\frac{2}{3}, \quad \frac{1}{2}+\frac{\sqrt{3}}{2}i, \quad -\frac{1}{2}+\frac{\sqrt{3}}{2}i, \quad \frac{1}{2}-\frac{\sqrt{3}}{2}i, \quad -\frac{1}{2}-\frac{\sqrt{3}}{2}i.$$

This verifies what we found. ∎

Finally, let's consider an example Newton himself gave.

Example 9.2.3. *Bound the number of positive and number of negative zeros of the polynomial*

$$P(x) = x^3 - 3a^2 x - 3a^3$$

for any real number $a > 0$.

Solution: First notice the a in this problem is a red herring. To see this consider instead the polynomial

$$Q(x) = \frac{1}{a^3} P(ax) = x^3 - 3x - 3.$$

If r is a zero of the Newton's polynomial, then r/a is a zero of our simplified polynomial—so they both have the same number of positive, negative and complex zeros. It does not matter what $a > 0$ is.

By Descartes' Rule of Sign, $Q(x)$ has only one positive zero, so either two or no negative zeros. So if Newton's Incomplete Rule indicates at least two complex zeros, then we are done! But sadly, the signs for the sequence in Newton's rule are $+$, $+$, $+$, $+$. So we gain no information here.

Now a little trick. Consider $Q(x-1) = x^3 - 3x^2 + 4x - 5$. This will have the same number of complex zeros as $Q(x)$ (and $P(x)$). For $Q(x-1)$, the sequence in Newton's Incomplete Rule is

$$1, \quad -\frac{1}{3}, \quad -\frac{29}{9}, \quad 25,$$

so there are at least two imaginary zeros. So we now know $Q(x)$ and $P(x)$ have one positive zero and two complex zeros. (The zeros of $P(x)$ are approximately $2.1038a$ and $(-1.0519 \pm 0.5652i)a$.) ∎

When reading this last example you might have asked "why did we consider $Q(x-1)$, not $Q(x+1)$ or $Q(x-7)$?" Trial and error. Newton put it as follows (quoted in [2]).

> And if there be any impossible roots it will rarely happen that they shall not be discovered in two or three such trials. Nor can there be an equation whose impossible roots may not be thus discovered.

9. SEPARATING THE REAL ZEROS OF POLYNOMIALS

By "impossible roots" Newton means complex zeros. Even though Newton saw negative zeros as "false" and imaginary zeros as "impossible," he did not hesitate to calculate them.

Exercises 9.2:

1. Use this section's two theorems to find the possible number of positive, negative and complex zeros for each of the following polynomials.

    a) $x^6 - 3x^3 + 2x^2 - 5$

    b) $x^4 + 8x^3 - 16x^2 - 52$

    c) $x^{60} - 3x^2 + 2x^2 - 1$

    d) $x^5 + x^4 + x^3 + 12x^3 - 12x^2 - 1$

    e) $x^5 + 3x^4 + 2x^2 + 11x + 3$

    f) $x^5 - 3x^4 - 2x^2 + 11x - 3$

2. Let $P(x) = a_0 x^n + a_1 x^{n-1} + \ldots + a_n$. Show that if $a_0 a_n > 0$ then P has an even number of positive zeros, and if $a_0 a_n < 0$ then P has an odd number of positive zeros.

3. (*) Prove the following.

    Lemma 9.2.3. *Let $P(x) = a_0 x^n + a_1 x^{n-1} + \ldots + a_n$ have real coefficients. Suppose there is some value of k for which $0 < k < n$ and $a_k^2 < a_{k-1} a_{k+1}$. Then P has at least one pair of complex zeros.*

    For example, the polynomials with real coefficients
    $$3x^{90} + \ldots - 5x^{45} + 4x^{44} - 10x^{43} + \ldots$$
    and
    $$3x^{45} + \ldots - x^{25} - 10x^{23} + \ldots$$
    each have at least one pair of complex zeros because
    $$4^2 - (-5)(-10) < 0 \quad \text{and} \quad 0^2 - (-1)(-10) < 0.$$

    Hint for proof: Use Newton's Incomplete Rule. Since the first and last term in his sequence is positive, it is sufficient to show the kth is negative. It will help to notice
    $$\left(\frac{k+1}{k}\right)\left(\frac{n-k+1}{n-k}\right) > 1.$$

4. Descartes, in his *La Geometrie* (1637), first proposed his rule for polynomials with non-zero coefficients (quoted in [2]). He stated it as follows.

We can determine also the number of true and false roots that an equation can have, as follows: An equation can have as many true roots as it contains changes of sign, from + to − or from − to +; and as many false roots as the number of times two + signs or two − signs are found in succession.

By "false roots" Newton means "negative zeros." Show that this rule is the same as Theorem 9.2.1. Is it still the same if we do not require that the coefficients be non-zero?

5. Look up the history of the numbers 0 and 1 (which were accepted as numbers long after 2, 3, 4, ...). Write a three page essay explaining why terms such as 'false' and 'impossible' were used in Newton's time period.

9.3 Bounds for the Zeros of Polynomials

If we are graphing a polynomial or separating its zeros, it helps to know roughly where the zeros are. One way to do this is to find a quick bound for the zeros. Cauchy's first bound below, for example, gives us the radius of a circle in the complex plane (with the origin as center), that contains all of the zeros (both real and complex) of the polynomial. We consider such bounds in this section. In the next we create similar bounds for just the real zeros. We will not give the sharpest possible bounds, but rather focus on those that are quick.

Theorem 9.3.1 (Cauchy Bound I). *Let $P(z) = z^n + a_1 z^{n-1} + \cdots + a_n$. The zeros of $P(z)$ all satisfy $|z| \leq 1 + \max(|a_1|, |a_2|, \ldots, |a_n|)$.*

(Here we used z not x to remind you that this bounds all of the zeros not just the real zeros.)

Proof. If 0 is a zero of $P(z)$, then it is bounded by any non-negative real number. So for the rest of the proof let's assume $z \neq 0$.

Divide $P(z)$ by $|z|^{n-1}$ to get

$$\frac{P(z)}{z^{n-1}} = z + a_1 + \frac{a_2}{z} + \frac{a_3}{z^2} + \cdots + \frac{a_n}{z^{n-1}}$$

and then use both forms of the triangle inequality (see page 14):

$$\left|\frac{P(z)}{z^{n-1}}\right| \geq |z| - \left|a_1 + \frac{a_2}{z} + \frac{a_3}{z^2} + \cdots + \frac{a_n}{z^{n-1}}\right|$$

$$\geq |z| - \left(|a_1| + \left|\frac{a_2}{z}\right| + \left|\frac{a_3}{z^2}\right| + \cdots + \left|\frac{a_n}{z^{n-1}}\right|\right).$$

9. SEPARATING THE REAL ZEROS OF POLYNOMIALS

Let $M = \max(|a_1|, |a_2|, \ldots, |a_n|)$, then $\dfrac{M}{|z|^j} \geq \left|\dfrac{a_{j+1}}{z^j}\right|$, so

$$\left|\frac{P(z)}{z^{n-1}}\right| \geq |z| - M\left(1 + \frac{1}{|z|} + \frac{1}{|z|^2} + \cdots + \frac{1}{|z|^{n-1}}\right)$$

$$\geq |z| - M\frac{\left(1 - \frac{1}{|z|^n}\right)}{\left(1 - \frac{1}{|z|}\right)} = |z| - M\frac{|z|^n - 1}{|z|^n - |z|^{n-1}}$$

so

$$|P(z)| \geq \frac{|z|^n(|z| - (1+M)) + M}{|z| - 1}.$$

If $|z| > 1 + M \geq 1$, then the right side is positive, so $P(z) \neq 0$ when $z \neq 0$. $\square$

The bound above can be slightly improved to

$$|z| \leq \max(|a_n|, 1 + |a_{n-1}|, \ldots, 1 + |a_1|, 1 + |a_0|). \tag{9.1}$$

A similar argument gives the following common form

$$|z| \leq \max(1, |a_1| + |a_2| + \ldots + |a_n|). \tag{9.2}$$

This bound is most useful when the absolute values of the coefficients are small.

The key step in the previous proof was to view the terms as a geometric series. The following theorem's proof uses that same idea in a different way.

Theorem 9.3.2 (Cauchy Bound II). *Let $P(z) = z^n + a_1 z^{n-1} + \cdots + a_n$. The zeros of $P(z)$ all satisfy $|z| \leq \max(|a_1|, |2a_2|^{1/2}, |3a_3|^{1/3}, \ldots, |na_n|^{1/n})$.*

Proof. Divide $P(z)$ by z^n and then begin as in the previous proof,

$$\left|\frac{P(z)}{z^n}\right| \geq 1 - \left(\left|\frac{a_1}{z}\right| + \left|\frac{a_2}{z^2}\right| + \left|\frac{a_3}{z^3}\right| + \cdots + \left|\frac{a_n}{z^n}\right|\right). \tag{9.3}$$

Now suppose $|z^j| < |ja_j|$ (for all $1 \leq j \leq n$), then $\left|\dfrac{a_j}{z^j}\right| \leq \dfrac{1}{n}$, and it follows,

$$\left|\frac{P(z)}{z^n}\right| \geq 1 - \left(\frac{1}{n} + \frac{1}{n} + \cdots + \frac{1}{n}\right).$$

The sum in parenthesis on the right side is less than one, so the right side is positive and $P(z) \neq 0$ when $z \neq 0$. $\square$

9.3. Bounds for the Zeros of Polynomials

Example 9.3.1. *Find a bound for the absolute values of the zeros of*

$$P(x) = z^9 + 3z^4 - 8z^3 + 1.$$

Solution: By Theorem 9.3.1, the zeros satisfy $|z| \leq 1 + \max(3, 8, 1) = 9$ (no reason to list the zero coefficients in the maximum). And by Theorem 9.3.2, the zeros also satisfy $|z| \leq \max(\sqrt[5]{15}, \sqrt[6]{48}, 1) \approx 1.906$. The zeros and this second bound are plotted on the left in Figure 9.1. ∎

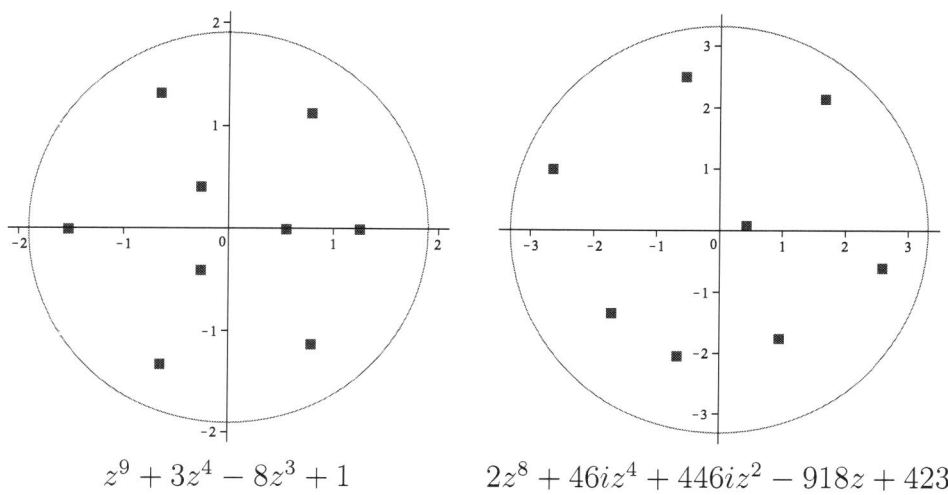

$z^9 + 3z^4 - 8z^3 + 1$ $\qquad\qquad 2z^8 + 46iz^4 + 446iz^2 - 918z + 423$

Figure 9.1: Zeros of Two Polynomials and Their Bounds

Example 9.3.2. *Find a bound for the absolute values of the zeros of*

$$P(x) = 2z^8 + 46iz^4 + 446iz^2 - 918z + 423.$$

Solution: Note that the theorems in this one section do not require that the coefficients are real. To use the theorems above, we must first divide the polynomial by 2 to get the following monic polynomial.

$$z^8 + 23iz^4 + 226iz^2 - 459z + \frac{423}{2}$$

Theorem 9.3.1 gives the bound $1 + \max(23, 226, 459, 423/2) = 460$. Theorem 9.3.2 gives the bound

$$\max(\sqrt[4]{4 \cdot 23}, \sqrt[6]{6 \cdot 226}, \sqrt[7]{7 \cdot 459}, \sqrt[8]{8 \cdot 423/2}) \approx 3.320.$$

The zeros and this second bound are plotted on the right in Figure 9.1. ∎

9. Separating the Real Zeros of Polynomials

Exercises 9.3:

1. Use Theorems 9.3.1 and 9.3.2 to find bounds for the zeros of the following.

    a) $x^4 + 12x^3 + 2x^2 + 5$

    b) $x^3 + 8x^2 - 16x - 12$

    c) $x^{40} - 3x^2 + 2x^2 - 1$

    d) $x^5 + x^4 + x^3 + 12x^3 - 12x^2 - 1$

    e) $x^5 + x^4 + x^3 + x^2 + x + 1$

    f) $5x^5 + 4x^4 + 3x^3 + 2x^2 + x + 1$

2. In both of our examples Theorem 9.3.2 gave smaller bounds than Theorem 9.3.1. Prove this is always the case. Hint: You might start by showing $|kx|^{1/k} < 1 + |x|$ for real numbers x and integers $k \geq 1$; then apply this result to the coefficients of the polynomials.

3. Prove the following theorem of Lagrange.

> **Theorem 9.3.3 (Lagrange Bound).** Let $P(z) = z^n + a_1 z^{n-1} + \cdots + a_n$. The zeros of $P(z)$ all satisfy $|z| \leq 2\max(|a_1|, |a_2|^{1/2}, |a_3|^{1/3}, \ldots, |a_n|^{1/n})$.

    Hint: You can follow the same structure as in the proof of Theorem 9.3.2. Use that if $|z| > 2|a_j|^{1/j}$ (for all $1 \leq j \leq n$), then $\left|\dfrac{a_j}{z^j}\right| \leq \left(\dfrac{1}{2}\right)^j$ in equation 9.3.

4. Show that Theorem 9.3.2 gives bounds that are always less than or equal to those from Theorem 9.3.3.

9.4 Bounds for Real Zeros

In this section we will restrict our study to polynomials with real coefficients.

When finding real zeros, we often look first for rational zeros using synthetic division—so it makes sense to start by showing how to find bounds on the zeros with synthetic division. Let us begin with an example.

Let $P(x) = x^4 - x^3 + 4x^2 + 16x - 8$. If we divide this by $x - 1$ we get the following.

```
1 | 1  -1   4   16   -8
  |     1   0    4   20
  -----------------------
    1   0   4   20 | 12
```

9.4. Bounds for Real Zeros

The remainder, and all of the coefficients of the quotient, are non-negative. Clearly, if we use any number larger than 1, these coefficients will just get more positive. So whatever the real zeros are, they are less than 1. Note that 1 is not a bound for the absolute values of the complex zeros (see the left image in Figure 9.2); so this bound for the real zeros is better than any possible bound on the complex zeros (like those found in the previous section).

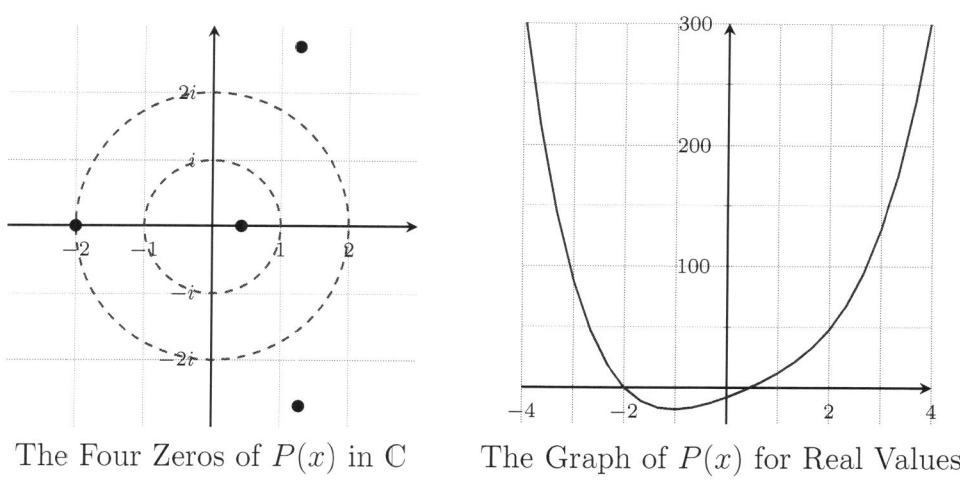

The Four Zeros of $P(x)$ in $\mathbb{C}$ — The Graph of $P(x)$ for Real Values

Figure 9.2: The Polynomial $P(x) = x^4 - x^3 + 4x^2 + 16x - 8$

For the negative zeros of $P(x) = x^4 - x^3 + 4x^2 + 16x - 8$, we could change the sign of the zeros by considering $P(-x) = x^4 + x^3 + 4x^2 - 16x - 8$. If we do so, we find that $x = 2$ is zero, and more apropos, a bound for the positive zeros of this polynomial.

$$
\begin{array}{r|rrrrr}
2 & 1 & 1 & 4 & -16 & -8 \\
  &   & 2 & 6 & 20 & 8 \\
\hline
  & 1 & 3 & 10 & 4 & 0
\end{array}
$$

This shows $x = -2$ is a lower bound for the negative zeros of $P(x)$.

This idea can be applied without making the change of sign as stated in the following theorem.

Theorem 9.4.1. *Let $P(x)$ be a polynomial with real coefficients.*

1. *If $P(x)$ is divided by $x - b$ ($b > 0$) using synthetic division, and the row containing the quotient and remainder has no changes of sign, then b is an upper bound for the real zeros of P.*

> 2. If $P(x)$ is divided by $x - b$ ($b < 0$) using synthetic division, and the row containing the quotient and remainder has entries that are alternating between positive and negative, then b is a lower bound for the real zeros of P.

The reasoning above can easily be reworded into a proof of this theorem.

There are a number of other ways to bound the zeros of a polynomial. In the next theorem we give three (and the exercises include several more).

> **Theorem 9.4.2 (Upper Bounds).** Let $P(x) = x^n + a_1 x^{n-1} + \ldots + a_n$. Upper bounds for the positive zeros of $P(x)$ are given by:
>
> 1. Choose $r > 0$ and let $-s$ be the sum of the negative quantities among
> $$a_1, \frac{a_2}{r}, \frac{a_3}{r^2}, \ldots \frac{a_n}{r^{n-1}},$$
> then $m = \max(r, s)$ is an upper bound.
>
> 2. If for each negative coefficient a_k we form the quotient
> $$\frac{-a_k}{\text{the sum of all the previous positive coefficients}} = \frac{-a_k}{\sum_{0 \leq j < k,\, a_j > 0} a_j},$$
> then one plus the largest of these quotients is an upper bound.
>
> 3. If the first negative coefficient is a_r and the least is a_k then $m = \sqrt[r]{-a_k} + 1$ is an upper bound.

To get lower bounds for the negative zeros of $P(x)$, consider the polynomial $P(-x)$, that is, change the sign of the coefficients of odd power terms and then use the bounds given above.

To get lower bounds for the positive zeros we consider $x^n P(1/x)/a_n$ (this inverts the zeros),[1] and then use the bounds above—inverting the resulting upper bounds to get lower bounds of the original polynomial.

Example 9.4.1. Let $P(x) = x^4 - 4x^3 - 4x^2 + 16x - 8$. Find bounds for the real zeros of $P(x)$.

---

[1] If a_n is zero, then $P(0) = 0$, so divide your polynomial by x. It makes no sense to try to bound the zeros of $P(x)$ when you already know one or more zeros. Divide the known zeros out, then bound what is left.

Solution: We use the methods above to find upper bounds for the zeros and get the second column of bounds in Table 9.1. For part (1) of the theorem, it can take a little trial and error to find the best choice of r. But since the first negative coefficient is $a_1 = -4$, it would not make sense to try an r less than 4 (or the sum in the theorem would automatically be larger that r). Trying $r = 4$ gives the bound

$$\max(4, 4 + 4/4 + 8/4^3) = 5.125.$$

Since increasing r might decrease this sum, we next try $r = 5$ and get

$$\max(5, 4 + 4/5 + 8/5^3) = 5.$$

At $r = 5$, r is the larger term and at $r = 4$, the sum was larger; so it is likely that in between 4 and 5 there is a better choice of r. (At the best choice of r, r will equal the sum.)

The other parts of Theorem 9.4.2 do not require trial and error. The bounds they give are listed in the third column of Table 9.1. The best of these upper bounds is 5.

Found Using	Lower Bound for Positive	Upper Bound for Positive	Upper Bound for Negative	Lower Bound for Negative
(1)	($r = 2$) 0.49	($r = 4$) 5.13	($r = 1$) -1.00	($r = 2$) -3.5
(1)		($r = 5$) 5.00	($r = .85$) -1.17	($r = 3$) **-3.00**
(2)	0.33	**5.00**	$-.875$	-4.20
(3)	**0.50**	8.00	**-1.26**	-4.00

Table 9.1: Bounds for the Zeros of $x^4 - 4x^3 - 4x^2 + 16x - 8$

To find a lower bound for the negative zeros we consider

$$P(-x) = x^4 + 4x^3 - 4x^2 - 16x - 8.$$

Since the zeros of this polynomial are the negatives of the zeros of $P(x)$, we need to find an upper bound for the zeros of $P(-x)$ and change the sign.

The first negative coefficient is $a_2 = -4$, so the sum when using part (1) of the theorem will be at least $4/r^2$; to get the minimum value we want the sum to be r, so we try a value of r roughly near the real solution to $r = 4/r^2$, which is $r = \sqrt[3]{4} \approx 1.587$. For $r = 2$ we get the sum 3.5 (so the bound is 3.5), larger than r, so we might try increasing the value of r. At $r = 3$, the sum is less than r, so we get the bound 3. We might be able to do better with something between 2 and 3, but it does not look like we will gain much, so we settle on 3 as an upper bound of the zeros of $P(-x)$ and record -3 as a lower bound for the zeros of $P(x)$ in the last column of results in Table 9.1. This bound is better than what we get with the other two parts of the theorem.

9. SEPARATING THE REAL ZEROS OF POLYNOMIALS

To find a lower bound for the positive zeros we form the polynomial whose zeros are the reciprocals of the zeros of the original polynomial,

$$x^4 P\left(\frac{1}{x}\right) = -8x^4 + 16x^3 - 4x^2 - 4x + 1.$$

To apply our theorem we need to make this monic by dividing by the leading coefficient (this does not alter the zeros).

$$P_2(x) = -\frac{x^4}{8} P\left(\frac{1}{x}\right) = x^4 - 2x^3 + 0.50x^2 + 0.50x - 0.125$$

Now the smallest positive zero of $P(x)$ is the largest zero of this polynomial; so we will find an upper bound on the positive zeros of $P_2(x)$ and then invert it.

Again choosing r for part (1) of the theorem takes some thought; again the a_1 coefficient of $P_2(x)$ is negative so we first try $r = a_1 = 2$. This gives us a sum of 2.015625, very close to r, so we are not going to try to beat it. This gives the lower bound $1/2.015625 \approx 0.496$ for the positive zeros of $P(x)$. In this manner we find the second column of results in Table 9.1. The best (or greatest) lower bound on the positive zeros is 0.50.

Finally, to find an upper bound for the negative zeros we form the polynomial whose zeros are the negative reciprocals of the zeros of the original polynomial:

$$P_3(x) = -\frac{x^4}{8} P\left(-\frac{1}{x}\right) = x^4 - 2x^3 + 0.50x^2 + 0.50x - 0.125.$$

We find an upper bound for the zeros of this polynomial, then invert and negate it. In this manner we find the fourth column of results in Table 9.1. The best (or smallest) upper bound for the negative zeros is -1.26.

The actual zeros are as follows and verify that bounds we found were reasonable.

$$\begin{aligned} 1 + \sqrt{2} + \sqrt{3} &\approx 4.14626 \\ 1 - \sqrt{2} + \sqrt{3} &\approx 1.31784 \\ 1 + \sqrt{2} - \sqrt{3} &\approx 0.682163 \\ 1 - \sqrt{2} - \sqrt{3} &\approx -2.14626 \end{aligned}$$

∎

Example 9.4.2. Let $P(x) = 3x^5 + 4x^4 - 26x^3 + 27x + 8$. *Find bounds for the zeros of $P(x)$.*

Solution: First, we divide through by 3 to make this polynomial a monic as required in the theorem. We again use the methods above to find bounds for the zeros. See the results in Table 9.2 and a graph in Figure 9.3. The actual zeros are (approximately) as follows.

$$-1/3, \quad -0.784705, \quad -3.57626, \quad 1.68048 \pm 0.163418\,i$$

Found Using	Lower Bound for Positive	Upper Bound for Positive	Upper Bound for Negative	Lower Bound for Negative
(1)	$(r=1)$ 0.30	$(r=1)$ 8.67	$(r=1)$ -0.26	$(r=1)$ -12.76
(1)	$(r=1.8)$ 0.55	$(r=2.9)$ 2.99	$(r=3)$ -0.29	$(r=3.7)$ -3.69
(2)	0.57	4.72	-0.23	-9.67
(3)	**0.67**	**2.94**	**-0.29**	**-3.38**

Table 9.2: Bounds for the Zeros of $3x^5 + 4x^4 - 26x^3 + 27x + 8$

Again the bound from part (1) from Theorem 9.4.2 does give some of the best bounds, but only after experimenting with the parameter r. We might as well experiment with c in synthetic division. In the last example synthetic division of $P(x)$ with $x = 2$ (i.e., upon division by $x - 2$), gives a quotient with all positive coefficients, so 2 is an upper bound for the positive zeros, and in fact is an upper bound that beats all of the bounds from Theorem 9.4.2. There just is no single method that is always best.

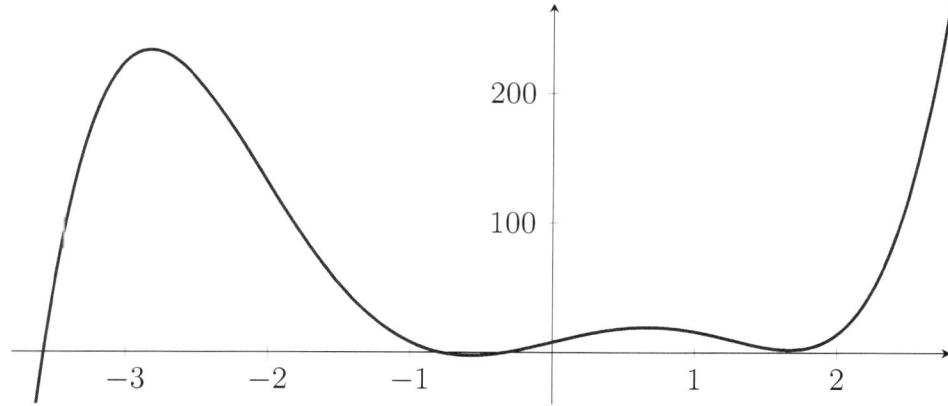

Figure 9.3: A Graph of $3x^5 + 4x^4 - 26x^3 + 27x + 8$

Exercises 9.4:

1. Use Theorem 9.4.1 and three parts of Theorem 9.4.2 to find upper bounds for the positive zeros of the following polynomials. Indicate which of these four bounds is best for each case.

9. SEPARATING THE REAL ZEROS OF POLYNOMIALS

   a) $x^6 - 3x^3 + 2x^2 - 5$

   b) $x^4 + 8x^3 - 16x^2 - 52$

   c) $x^{60} - 3x^2 + 2x^2 - 1$

   d) $x^5 + x^4 + x^3 + 12x^3 - 12x^2 - 1$

   e) $x^5 + 3x^4 + 2x^2 + 11x + 3$

   f) $x^5 - 3x^4 - 2x^2 + 11x - 3$

2. Use Theorem 9.3.2 and all parts of Theorem 9.4.2 to find lower bounds for the positive zeros of the polynomials in problem 1. Indicate which of these four bounds is best for each case.

3. Use Theorem 9.4.1 and all parts of Theorem 9.4.2 to find lower bounds for the negative zeros of the polynomials in problem 1. Indicate which of these four bounds is best for each case.

4. Use Theorem 9.3.2 and all parts of Theorem 9.4.2 to find upper bounds for the negative zeros of the polynomials in problem 1. Indicate which of these four bounds is best for each case.

5. Give an example to show that Theorem 9.3.1 can give a better bound than Theorem 9.4.2.

6. (*) Prove Theorem 9.4.2 part (1).

7. (*) Theorem 9.4.2 part (2).

8. Let $P(x) = x^n + a_1 x^{n-1} + \ldots + a_n$. Prove that if $-c \leq a_i$ for all i, then $m = 1 + c$ is an upper bound for the real zeros of $P(x)$. Hint: Use synthetic division. Show that if c is chosen as in the theorem, then every coefficient of the quotient, and the remainder, are at least one.

9. Let $P(x) = x^n + a_1 x^{n-1} + \ldots + a_n$. Prove that if $a_1, a_2, \ldots, a_r$ are all non-negative and $-c \leq a_i$ for all i, then

$$m = 1 + \frac{c}{1 + a_1 + a_2 + \ldots + a_r}$$

is an upper bound. Hint: Theorem 9.4.2 part (2).

10. (*) Prove that if m is any positive number such that the polynomial values $P(m)$, $P'(m)$, $P''(m)$, $\ldots$, $P^{(n)}(m)$ are all positive, then m is an upper bound for the real zeros of $P(x)$. Hint: Induction.

9.5 Sturm's Theorem

In this section we will look at a method that tells us exactly how many zeros of $P(x)$ are in any given interval (a, b). To do this we first need to define a sequence of polynomials using a method that is essentially the Euclidean algorithm from section 3.4. Given any polynomial $P(x)$ we define its **Sturm sequence** to be the sequence of polynomials

$$P_0 = P, \quad P_1 = P', \quad P_2, \quad P_3, \quad \ldots, \quad P_k,$$

where P_k is a greatest common divisor of P and P'; and where

$$-P_{i+1} = \text{rem}(P_{i-1}, P_i) \qquad (1 < i < k)$$

is the usual remainder from the Euclidean algorithm with a negative sign. (We will work out a detailed example below.)

Now let **Var (P, a)** be the number of variations in sign in the Sturm sequence of $P(x)$ evaluated at the real number a. This function is key to finding the number of zeros in an interval as shown in the following theorem.

Theorem 9.5.1 (Sturm's Theorem). *Let $P(x)$ be a polynomial and let (a, b) be any interval of real numbers. The number of distinct zeros of $P(x)$ in the interval $(a, b]$ is exactly the difference*

$$\text{Var}(P, a) - \text{Var}(P, b).$$

(The proof is in Appendix A.2.)

Let's first illustrate this with a simplistic example. This should help you understand the process.

Example 9.5.1. *Let $P(x) = x^4 - 2x^3 + 2x^2 - 2x + 1$. Find $\text{Var}(P, 0)$, $\text{Var}(P, \infty)$ and $\text{Var}(P, -\infty)$. What do these tell us?*

Solution: The sequence starts with

$$P_0(x) = P(x) = x^4 - 2x^3 + 2x^2 - 2x + 1$$

and

$$P_1(x) = P'(x) = 4x^3 - 6x^2 + 4x - 2.$$

9. SEPARATING THE REAL ZEROS OF POLYNOMIALS

To find $P_2(x)$, we divide $P_0(x)$ by $P_1(x)$ and keep just the remainder.

$$
\begin{array}{r}
\frac{1}{4}x - \frac{1}{8} \\
4x^3 - 6x^2 + 4x - 2 \overline{)\, x^4 - 2x^3 + 2x^2 - 2x + 1} \\
-x^4 + \frac{3}{2}x^3 - x^2 + \frac{1}{2}x \\
\overline{-\frac{1}{2}x^3 + x^2 - \frac{3}{2}x + 1} \\
\frac{1}{2}x^3 - \frac{3}{4}x^2 + \frac{1}{2}x - \frac{1}{4} \\
\overline{\frac{1}{4}x^2 - x + \frac{3}{4}}
\end{array}
$$

After the required sign change we see

$$P_2(x) = -\tfrac{1}{4}x^2 + x - \tfrac{3}{4}.$$

Next we divide $P_1(x)$ by $P_2(x)$ to find the remainder $-P_3(x)$.

$$
\begin{array}{r}
-16x - 40 \\
-\tfrac{1}{4}x^2 + x - \tfrac{3}{4} \overline{)\, 4x^3 - 6x^2 + 4x - 2} \\
-4x^3 + 16x^2 - 12x \\
\overline{ 10x^2 - 8x - 2} \\
-10x^2 + 40x - 30 \\
\overline{ 32x - 32}
\end{array}
$$

This shows $P_3(x) = -32x + 32$. Finally we see $P_3(x)$ exactly divides $P_2(x)$, so the Sturm sequence is

$$
\begin{aligned}
P_0(x) &= x^4 - 2x^3 + 2x^2 - 2x + 1, \\
P_1(x) &= 4x^3 - 6x^2 + 4x - 2, \\
P_2(x) &= -\tfrac{1}{4}x^2 + x - \tfrac{3}{4}, \\
P_3(x) &= -32x + 32, \\
P_4(x) &= 0.
\end{aligned}
$$

Since 0 is ignored when counting changes of sign we can omit $P_4(x)$ in the sign counting below.

To find $\mathrm{Var}(P, 0)$, we substitute $x = 0$ into the terms of this sequence and we get

$$1, \quad -2, \quad -\tfrac{3}{4}, \quad 32.$$

This shows two changes of sign, so $\mathrm{Var}(P, 0) = 2$. To find $\mathrm{Var}(P, \infty)$, we just ask what the sign of the terms are when we substitute in arbitrarily large values of x:

$$+, \quad +, \quad -, \quad -.$$

9.5. Sturm's Theorem

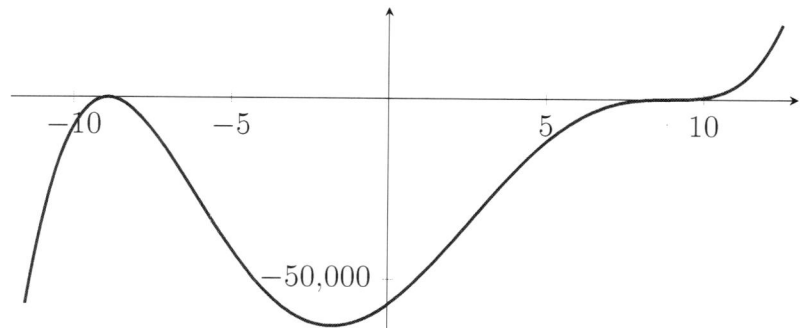

Figure 9.4: Graphs of $x^5 - 9x^4 - 159x^3 + 1431x^2 + 6320x - 56885$

Similarly for $\operatorname{Var}(P, -\infty)$, we find the signs for arbitrarily negative values of x:

$$+, \underbrace{-,}_{\uparrow} \underbrace{-,}_{\uparrow} +.$$

This shows $\operatorname{Var}(P, \infty) = 1$ and $\operatorname{Var}(P, -\infty) = 2$. So these tell us the number of real zeros in $(-\infty, \infty)$ is exactly $2 - 1 = 1$, and using $\operatorname{Var}(P, 0) = 2$, we see this one real zero is positive. ∎

In that last example we know the one real zero because the last non-zero remainder in this process, $P_3(x) = -32(x-1)$, is always a gcd of $P(x)$ and $P'(x)$. So $x = 1$ is a double zero of $P(x)$. Using this, we find

$$P(x) = (x-1)^2(x^2+1).$$

This means the one real zero is the double zero $x = 1$. Note that unlike Newton's method, Sturm's theorem counts distinct zeros, not their multiplicity.

Example 9.5.2. *Separate the real zeros of*

$$P(x) = x^5 - 9x^4 - 159x^3 + 1431x^2 + 6320x - 56885.$$

Solution: By Theorem 9.3.2 the zeros (real and complex) are in a circle of radius less than 13 about the origin. By Descartes' Rule of Sign, this polynomial has three or one positive zeros and two or zero negative zeros (counting multiplicity). A quick graph seems to suggest the same (see Figure 9.4). Now we perform the necessary division to find the Sturm sequence.

$$\begin{aligned}
P(x) &= x^5 - 9x^4 - 159x^3 + 1431x^2 + 6320x - 56885 \\
P'(x) &= 5x^4 - 36x^3 - 477x^2 + 2862x + 6320 \\
P_2(x) &= \tfrac{1914}{25}x^3 - \tfrac{17172}{25}x^2 - \tfrac{152158}{25}x + \tfrac{273049}{5} \\
P_3(x) &= \tfrac{9950}{305283}x^2 + \tfrac{127375}{610566}x - \tfrac{99125}{101761} \\
P_4(x) &= -\tfrac{29582024461}{7920200}x - \tfrac{15393081187}{792020} \\
P_5(x) &= \tfrac{10122865378058500}{8599524092837161}
\end{aligned}$$

195

9. Separating the Real Zeros of Polynomials

We see $\text{Var}(P, \infty)$ is just the number of changes in sign on the leading terms $(+ + + + - +)$, so is two. Next $\text{Var}(P, -\infty)$ can also be found from the leading coefficients viewing x as negative $(- + - + + +)$, so we get three changes. We now know that there are exactly $3-2=1$ real zeros. From the graph (Figure 9.4), it must be positive and near 10. To nail it down more closely, we evaluate the Sturm sequence at a few different values.

x	Sturm Sequence of P evaluated at x	$\text{Var}(P, x)$
8	-245, 736, 1157.6, 2.78078, -49315.2, 1.177	3
9	-5, 2, 7.8800, 3.54347, -53050.3, 1.177	3
10	415, 1240, 1618.6, 4.37135, -56785.3, 1.177	2

We see the one real zero is in the interval $(9, 10)$. With this extra information we can zoom in on a couple places in the graph (Figure 9.5). ∎

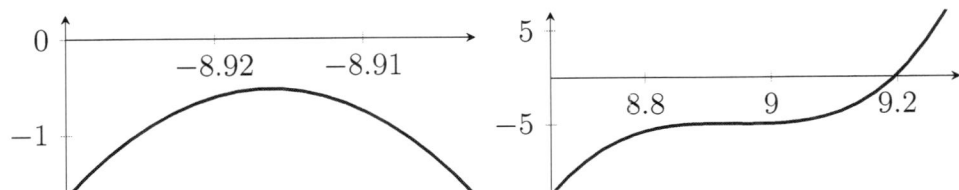

Figure 9.5: Two Graphs of $x^5 - 9x^4 - 159x^3 + 1431x^2 + 6320x - 56885$

Using Computers

While looking at the last example you might have been surprised (dismayed?) at the size of the fractions in the coefficients of the Sturm sequence. We could simplify this process slightly by dividing through by the absolute value of the leading coefficient every time we find a polynomial in Sturm's sequence. But this helps only a little. Another option is to use a computer.

Maple has the commands `sturmseq(poly,x)` which finds a normalized Sturm sequence. Maple also has `sturm(SturmSeq,x,a,b)` which is the number of distinct real zeros in the interval $(a, b]$. Before calculating the Sturm sequence, Maple divides $P(x)$ by $\gcd(P(x), P'(x))$ so all multiple zeros become simple zeros.

Exercises 9.5:

1. Use Sturm's theorem to separate the zeros of the following polynomials.

    a) $x^3 - 3x + 1$

    b) $x^3 - 5x + 5$

    c) $x^n + x$ (for any integer $n > 0$)

d) $x^3 + x^2 + x - 1$

e) $x^3 - x^2 - 2x + 1$

f) $x^4 - 2x^2 + x + 1$

g) $x^4 + x^3 - 2x^2 - x$

h) $x^8 + x^7 + 1$

i) $1 + x + \dfrac{x^2}{2} + \dfrac{x^3}{3} + \dfrac{x^4}{4}$

2. (*) Prove that a polynomial of degree n has n distinct real zeros if and only if its Sturm sequence consists of n terms whose leading coefficients all have the same sign.

3. (**) Suppose the Sturm sequence for $P(x)$ begins $S_1, S_2, \ldots, S_k$ where the last term is never zero (hence never changes sign), in the interval $[a, b]$. Show that we may use this shortened sequence to calculate the number of zeros. Hint: How many changes of sign can there be in $S_k, S_{k+1}, \ldots$?

4. Let n be a positive integer and let

$$P(x) = 1 + x + \frac{x^2}{2!} + \frac{x^3}{3!} + \cdots + \frac{x^n}{n!}.$$

a) Show that $P(x), P'(x), -x^n$ will work as a (shortened) Sturm sequence for $P(x)$ on the intervals $(-\infty, -\epsilon)$ and (ϵ, ∞) for an $\epsilon > 0$ (see the previous problem). Hint: $P(x) - P'(x) = x^n/n!$. (This problem is from [30, Chapter Four].)

b) Use part (a) to show that $P(x)$ has exactly one real zero when n is odd and none when n is even.

Approximating Real Zeros: Numerical Methods

He who seeks for methods without having a definite problem in mind seeks for the most part in vain.

D. Hilbert

Corporal Nym: I have operations in my head, which be humorous of revenge.

William Shakespeare

Euler calculated without apparent effort, just as men breathe, as eagles sustain themselves in the air.

Arago

Chapter 10

Approximating Real Zeros: Numerical Methods

In this chapter we outline several techniques for finding the real zeros (roots) of equations in general and of polynomial equations in particular. These techniques are probably programmed into your computer and calculator, and perhaps soon your watch, but our goal here is to understand how these processes work. We hope to expand our own understanding so that we can master technology rather than be mastered by it.

Most of these methods can be adapted to find complex zeros, and we discuss this in some of the exercises.

10.1 Sure and Simple: Bisection

In the previous chapter, "Separating the zeros of polynomials," we saw how to isolate each real zero of a polynomial $P(x)$ in an interval (a, b). If we can make the interval small enough, then any point in the interval will be a good approximation of the zero. The method of bisection does this by repeatedly bisecting the interval, and deciding in which half the zero lies. The key to deciding where the zero lies is the following theorem version of the Intermediate Value Theorem.

Theorem 10.1.1 (Weierstrass' Nullstellensatz). *Let $f(x)$ be a function that is continuous on the closed interval $[a, b]$. If $f(a)f(b) < 0$, then there exists at least one zero of $f(x)$ between a and b.*

Example 10.1.1. *Approximate the zeros of $f(x) = x^3 - 6x + 2$.*

10. APPROXIMATING REAL ZEROS: NUMERICAL METHODS

Solution: All polynomials are continuous, so by checking the value of $f(x)$ at the endpoints we can see there are zeros in each of the following intervals: $(-3, -2)$, $(0,1)$ and $(2,3)$. We first find the zero in the interval $(0,1)$.

We know there is a zero in $(0,1)$ because $f(0) = 2$, $f(1) = -3$. If we use the midpoint of this interval 0.5 as an estimate for the zero, the error is less than one half of the width of the interval (also 0.5). To improve our estimate calculate the value of f at the midpoint: $f(0.5) = -0.875$. Because $f(0)$ and $f(0.5)$ have opposite signs, we know the zero is in the left subinterval $(0, 0.5)$. We now check the midpoint 0.25 of this new interval: $f(0.25) = 0.515625$, finding the zero is in the right subinterval $(0.25, 0.5)$. Now if we use the midpoint as an estimate, the error is at most 0.125, again one half of the width of the interval. We continue bisecting the interval in which the zero lies, each time dividing the maximum error in our approximation by two. We record our results in Table 10.1.

step	left endpoint	right endpoint	f(midpoint)	width of interval
0	0.0000000	1.0000000	-0.8750000	1.0000000
1	0.0000000	0.5000000	0.5156250	0.5000000
2	0.2500000	0.5000000	-0.1972656	0.2500000
3	0.2500000	0.3750000	0.1555176	0.1250000
4	0.3125000	0.3750000	-0.0218811	0.0625000
5	0.3125000	0.3437500	0.0665779	0.0312500
6	0.3281250	0.3437500	0.0222869	0.0156250
7	0.3359375	0.3437500	0.0001873	0.0078125
8	0.3398438	0.3437500	-0.0108509	0.0039063
9	0.3398438	0.3417969	-0.0053327	0.0019531
10	0.3398438	0.3408203	-0.0025730	0.0009765
11	0.3398438	0.3403320	-0.0011928	0.0004883
12	0.3398438	0.3400879	-0.0005028	0.0002442
13	0.3398438	0.3399658	-0.0001578	0.0001221

Table 10.1: Bisection (Example 10.1.1)

If we choose the midpoint of the 13th interval (0.3399048) as our estimate of the zero we know the error is at most 1/2 of the width of that interval (0.0000611). If we continue bisecting these intervals another 40 times we find that one zero is 0.3398768866231826 with an error of at most 0.0000000000000001. Similarly, we can find the other two zeros with the same margin for error:

$$r_1 = 0.3398768866231826,$$
$$r_2 = -2.6016791311883154,$$
$$r_3 = 2.2618022452599720.$$

∎

The method we used in the last example is called bisection because each time we evaluated the function we bisected the the interval containing the zero. In Algorithm 10.1.1 bisection is presented as 'pseudocode' (an informal description of the code similar to an actual program). The 'while loop' that runs from line 2 to 9 is repeated as long as the interval's width is wider than ϵ. In this loop we first set M to be the midpoint of the interval (on line 3 the $\leftarrow$ is the 'assignment' operator, so M is assigned the value $(L+R)/2$).

Algorithm 10.1.1 Bisection Algorithm

Require: f continuous on $[L, R]$, $f(L)f(R) < 0$ and $\epsilon > 0$.

1: **procedure** BISECTION(f,L,R,ϵ)
2:     **while** $|R - L| > \epsilon$ **do**
3:         $M \leftarrow (R + L)/2$ ▷ Calculate the midpoint.
4:         **if** $f(L)f(M) > 0$ **then**
5:             $L \leftarrow M$ ▷ Keep the right half.
6:         **else**
7:             $R \leftarrow M$ ▷ Keep the left half.
8:         **end if**
9:     **end while**
10:     **return** M ▷ M is an approximate zero
11: **end procedure**

The if condition in this loop (line 4) always keeps the half of the interval which has a sign chances (hence a zero). Eventual the width of the interval is at most ϵ, so we exit the while loop and return the approximate zero M to the user.

One nice thing about bisection is that it is sure. At every step we have an explicit error bound, and in every step we know that bound is divided by two. Another strong point is that it can be used with any continuous function, not just polynomials. A third good point is that it is very simple, easy to use and to program. On the other hand, it has the disadvantage that we must begin with an interval for which the function has different signs at the end points, and this interval may be very hard to find. A second disadvantage is that it converges rather slowly—the algorithms we present later often find the zero far faster.

Example 10.1.2. *Write a computer program to implement the bisection algorithm.*

Solution: The algorithm above is easily translated into a program. We have decided to make the calculation of the function f a separate routine so that it can be easily changed. In `Maple` the program might be as follows.

```
bisection := proc(f, left, right, epsilon)
   local M, L, R;
   L := left; R := right;
   while epsilon < abs(R-L) do
      M := (L+R)/2;
      if f(L)*f(M) > epsilon then
         L := M
      else
         R := M
      end if;
   end do;
   return M;
end proc;
```

The first line tells Maple you are defining a procedure (code to implement an algorithm) and what information (parameters) you will give it. Maple uses := as an assignment operator (like ←). Notice that Maple will not let you alter the program parameters (f, left, right, epsilon) within the bisection procedure, so we create local variables (variables which only have meaning inside this one procedure) R and L and give them the initial values left and right respectively.

In Pascal we might write the following.

```
FUNCTION bisection (a, b, Error: REAL);
VAR L, R, M: REAL;
BEGIN
   L := A;
   R := B;
   WHILE (abs(R-L) > Error) DO
      BEGIN
         M := (R+L)/2
         IF f(M)f(L) > 0 THEN L = M ELSE R = M;
      END;
   bisection := M
END;
```

The point is that this can be easily implemented in any language, but each language has it its own foibles. That is one of the advantages of pseudocode—the details can be left to the implementor. ∎

Clearly much could be said about implementing the bisection method in any language, for example how using $f(M)f(L)$ might cause an avoidable underflow. We begin to address a couple such issues in the exercises, but we must leave in-depth understanding for numerical analysis and programming classes.

Exercises 10.1:

1. Find a zero of the following functions using the method of bisection.

   a) $x^2 - \cos x = 0$
   b) $x + 1 + \sin x = 0$
   c) $xe^x - 3 = 0$
   d) $x \log x = 15$
   e) $\sin x - \csc x = -1$
   f) $x^2 - 7$
   g) $x^3 - 56x + 1$
   h) $x^3 + 3x^2 - 4x - 12$
   i) $x^4 - 6x^3 + 12x^2 - 20x - 9$
   j) $x^5 + 6x^4 + 10x^2 + 3$

2. Run the following program (in traditional computer language of your choice). This program sets the variable *eps* to 1 and then repeatedly divides *eps* in half until $1 + eps \leq 1$. Mathematically this never will happen (because $1 + eps$ is always greater than 1) so why does the program stop?

   $eps \leftarrow 1$
   while $1 + eps > 1$ **do**
      $eps \leftarrow eps/2$
   end while

   (If you use a symbolic program like Maple, make sure it evaluates $1 + eps$ numerically by using evalf(1+eps) > 1.)

3. Run the program in the previous problem, then change $1+eps > eps$ to $1000+eps > 1000$ and run it again (try other values besides 1 and 1000). What can you conclude about how the maximum error should be chosen in a zero approximating program?

4. In you favorite language give an example of values for f, L, R and ϵ where Algorithm 10.1.1 would fail by being caught in an infinite loop. Hints: First consider the previous two problems. Second, some programs would replace line three with $M \leftarrow L + (R - L)/2$; do you see why?

5. Why use bisection instead of trisection? More generally, why not divide the interval into n subintervals? Hint: Your answer might include the words 'simplicity' and 'speed' (and must be in complete sentences.)

6. Show that Weierstrass' Nullstellenstaz is equivalent to the Intermediate Value theorem. That is, assuming the Intermediate Value Theorem prove the Nullstellenstaz, and then assuming the Nullstellenstaz prove Intermediate Value Theorem. You may take the following for the statement of the Intermediate Value Theorem.

> **Theorem 10.1.2 (Intermediate Value Theorem).** *Let f be a function that is continuous on the closed interval $[a, b]$. If $f(a) \neq f(b)$, then for every number k between $f(a)$ and $f(b)$ there exists at least one number c between a and b for which $f(c) = k$.*

10. Approximating Real Zeros: Numerical Methods

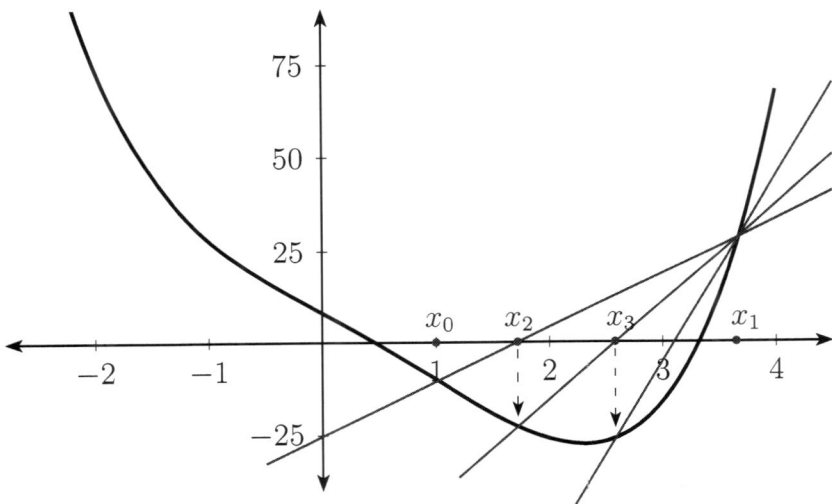

Figure 10.1: The Method of False Position (Regula Falsi)

7. (*R) Prove the Weierstrass' Nullstellenstaz. Hint: Apply the bisection algorithm to the interval $[a, b]$. Let the midpoints of the successive intervals be m_i ($i = 0, 1, 2, \ldots$). You need only show that the m_i converge to a point in the interval $[a, b]$, so you need only show the sequence $m_0, m_1, \ldots$ is a Cauchy sequence.

8. Rather than bisect the interval (L, R) we might draw the line through the points $(L, f(L))$ and $(R, f(R))$ and let M be the intersection of this line and the x-axis (see Figure 10.1). Show this is equivalent to line 3 in the Bisection Algorithm 10.1.1 with

$$M = \frac{f(R)L - f(L)R}{f(R) - f(L)}.$$

This is call the method of **False Position** or **Regula Falsi** which is often faster (but sometimes slower) than the method of bisection.

9. Repeat problem 1 using the method of Regula Falsi (see the previous problem).

10. (*) One of the problems with the method of bisection is that it may only be used to find real zeros—but recall that every polynomial with rational coefficients may be factored into the product of linear and quadratic polynomials with real coefficients. Once factored thus, it is a simple matter to find the zeros using the quadratic formula. As an example suppose $P(x)$ is a quartic and

$$P(x) = x^4 + ax^3 + bx^2 + cx + d = (x^2 + px + q)(x^2 + rx + s).$$

Show that the following relations hold:

$$s = \frac{d}{q}, \qquad r = \frac{c - as}{q - s}, \qquad p = a - r, \qquad s + pr + q - b = 0$$

206

Thus given q, we may find s, r and p in terms of q (and the coefficients of P). Show that if we do so and solve $s + pr + q - b = 0$ for q by bisection, then we may easily find the factorization, hence the zeros, of P (whether or not they are real).

11. 🖳 Write a computer program to implement the method suggested in the previous problem.

10.2 Faster But Not Sure: Newton-Raphson

Let x_0 be an initial estimate (or guess) for a solution of $f(x) = 0$. If $f(x)$ has a derivative at x_0, then near x_0, the function $f(x)$ is approximately equal to the tangent line $y = f'(x_0)(x - x_0) + f(x_0)$. So perhaps where the tangent line crosses the axis is a better estimate of the zero than x_0. See Figure 10.2 where we start with the initial guess $x_0 = 1$ and the tangent line crosses at $x = 1.2$, a little closer to the actual zero.

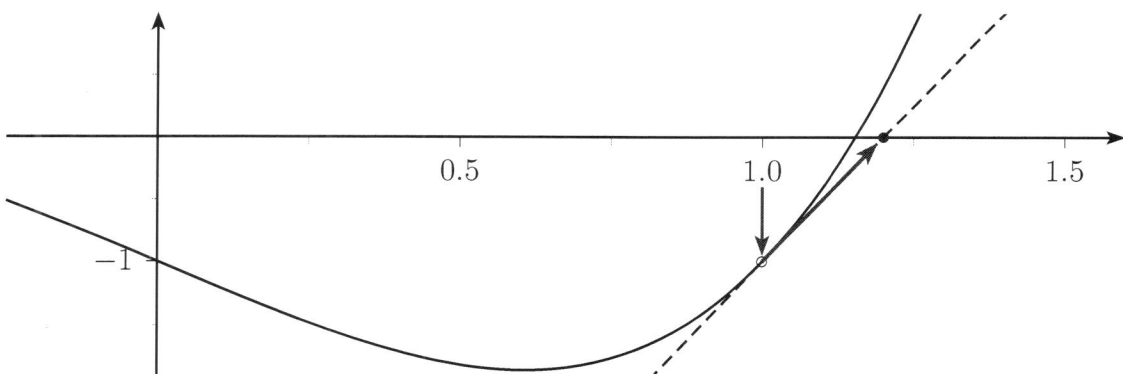

Figure 10.2: A Graph of $x^4 + x^3 - 2x - 1$

Newton's idea was to do this repeatedly defining a sequence of points which (hopefully) converge to the zero by

$$x_{i+1} = x_i - \frac{f(x_i)}{f'(x_i)} \qquad (10.1)$$

(see problem 10.1). This is know as **Newton's Method** or the **Newton-Raphson** method. For the example in Figure 10.2 we get the values in the following table.

10. APPROXIMATING REAL ZEROS: NUMERICAL METHODS

i	x_i
0	1.000000000000000000000000000000000000000000000000
1	1.200000000000000000000000000000000000000000000000
2	1.156499133448873483535528596187175043327556 3258232
3	1.153732177145488415644195990642244441601334 8331876
4	1.153721375705930094905029530145824714896939 8449219
5	1.153721375541767900903911284833409466914579 1505851
6	1.153721375541767900865992748763864001777261 0469639
7	1.153721375541767900865992748763864001775237 9886997
8	1.153721375541767900865992748763864001775237 9886997

Table 10.2: Calculations for $x^4 + x^3 - 2x - 1$ Starting with $x_0 = 1$

As you can see, when this method works, it tends to work very well (the digits of accuracy often double each application of equation 10.1). But there are many ways for this algorithm to fail: $f(x_i)$ or $f'(x_i)$ could be undefined; $f'(x_i)$ could be zero (or too close to it); or the process could produce a sequence $x_0, x_1, \ldots$ that does not converge. For example, in Figure 10.3, the first step from x_0 to x_1 is a large one, but then the next few seem to be approaching a value near 2, but there is no zero there.

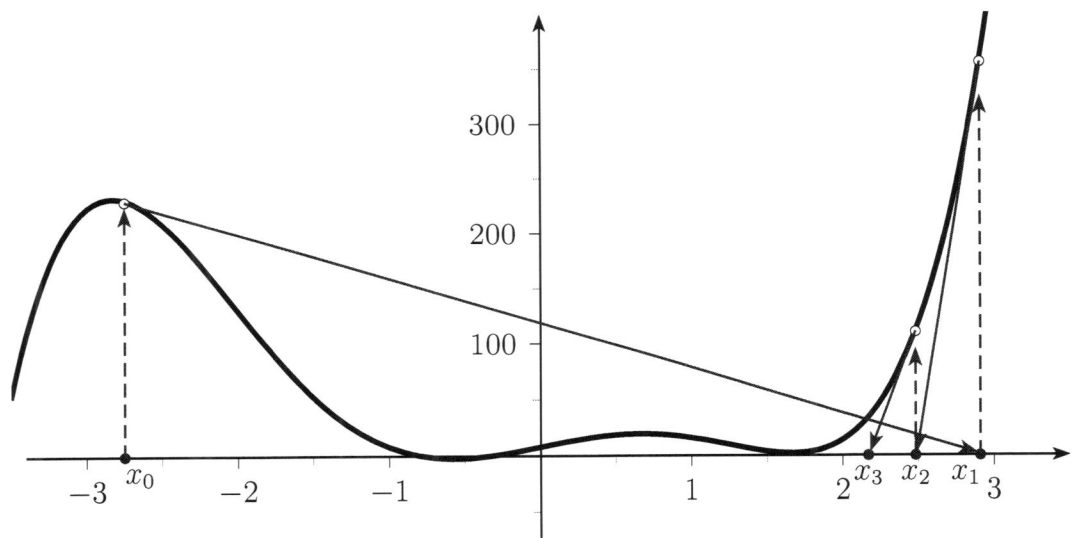

Figure 10.3: A Graph of $3x^5 + 4x^4 - 26x^3 + 27x + 8$

Example 10.2.1. *Approximate the zeros of $x^3 - 6x + 2$.*

Solution: Note first that $f'(x) = 3x^2 - 6$. We record the work in the Table 10.3, beginning with the initial values $x_0 = -3$, 0, and 2 (the left endpoints used in

10.2. Faster But Not Sure: Newton-Raphson

Example 10.1.1 of the previous section). Notice how quickly the values x_i converge to the zero.

x_0	i	x_i	$f(x_i)$	$f'(x_i)$
-3	0	-3.000000000000000	-7.000000000000000	21.00000000000000
	1	-2.666666666666666	-0.962962962962963	15.33333333333333
	2	-2.603864734299517	-0.031304965658405400	14.34033466358608
	3	-2.601681733247503	-0.000037215699048123	14.30624352334119
	4	-2.601679131886846	-0.000000000052817528	14.30620291588648
	5	-2.601679131883154	-0.000000000000000055	14.30620291582885
	6	-2.601679131883154	-0.000000000000000055	14.30620291582885
0	0	0.0000000000000000	2.000000000000000	-6.000000000000000
	1	0.3333333333333333	0.037037037037037	-5.666666666666666
	2	0.3398692810457516	0.000042997819117124	-5.653466615404332
	3	0.3398768866127501	0.000000000058979294	-5.653451105839272
	4	0.3398768866231826	0.000000000000000000	-5.653451105817997
	5	0.3398768866231826	0.000000000000000000	-5.653451105817997
2	0	2.000000000000000	-2.000000000000000	6.000000000000000
	1	2.333333333333333	0.703703703703704	10.33333333333333
	2	2.265232974910394	0.032147785493625824	9.393841291864185
	3	2.261810755554669	0.000079548328143108	9.347363681828352
	4	2.261802245312546	0.000000000491428869	9.347248190702630
	5	2.261802245259972	0.000000000000000000	9.347248189989147
	6	2.261802245259972	0.000000000000000000	9.347248189989147

Table 10.3: Calculations for Example 10.2.1

∎

Example 10.2.2. *Apply Newton's Method to $x^3 - 2x + 2$ with x_0 near 0 or 1.*

Solution: If we start with $x_0 = 0$, then

$$x_1 = 0 - \frac{2}{-2} = 1,$$

and

$$x_2 = 1 - \frac{1}{1} = 0,$$

so the sequence of values given by Newton's Method is the "2-cycle" $0, 1, 0, 1, 0, 1, \ldots$.

10. Approximating Real Zeros: Numerical Methods

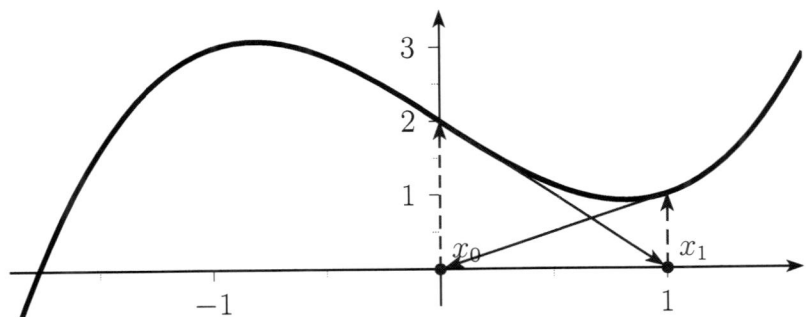

Figure 10.4: A "2-cycle" from the Graph of $x^3 - 2x^2 + 2$ for Example 10.2.2

We get caught in the same cycle if we start too near 0 or 1. For example, starting with $x_0 = 0.99$ we soon settle in on this same 2-cycle (see Table 10.4 and Figure 10.4).

i	x_i
0	0.980000000000000000000000000000000000000000000000
1	$-0.063173455280229713921089014144421992980963522280$
2	1.006276016496094514635469931083418262055662899672$8$
3	0.036513647513824698760545792598161214393984982148$7$
4	1.001955097932127539476220791066813972410570311166$1$
5	0.011617128032505728936844111676657307043333760138$3$
6	1.000200909346511038431013952141435576501114012895$4$
7	0.001204246470623197034438837669073892324000998258$7$
8	1.000002173572661469713252222801950262763534984247$94$
9	0.000013041294237958912936522757432178357753206251$7$
10	1.000000002551108151638601992060221463781678885934
11	0.000000001530664889030715357551947858585036957158$8$
12	1.000000000000000003514402500180869906300673336740$9$
13	0.000000000000000021086415001085219067273292022118$8$
14	1.000000000000000000000000000000000006669553463969876
15	0.000000000000000000000000000000000040017320783819$25$
16	1.000000000000000000000000000000000000000000000000
17	0.000000000000000000000000000000000000000000000000
18	1.000000000000000000000000000000000000000000000000

Table 10.4: Calculations for Example 10.2.2

■

Let's take a brief side step for a moment and think about this algorithm in the complex plane (instead of just the real numbers). If we use the same algorithm, but with complex numbers, we can find all of the zeros of $x^3 - 2x + 2$. In Figure 10.6 we

10.2. Faster But Not Sure: Newton-Raphson

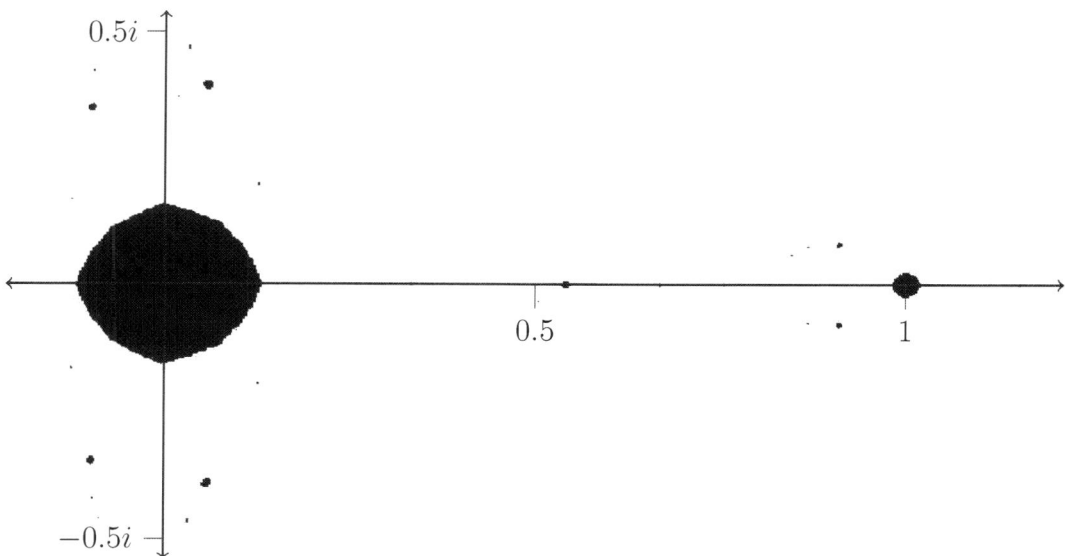

Figure 10.5: Regions where Newton's Method Fails to Converge for $z^3 - 2z + 2$

use the same polynomial $x^3 - 2x^2 + 2$ and shade the region based on how quickly the method converges to one of the zeros. If you color these (hard to do in a black and white text) using colors to show which of the three zeros the process converges to, you get amazing fractal images. More importantly, using complex numbers also makes it easier to visualize the problem we encountered in example 10.2. In Figure 10.5 we shade just those regions that lead to the 2-cycle discussed. You could do a Google search on "Newton Fractal" to find many examples of this type of image—creating such images yourself could be a fun research project.

We are now done with our side foray into the complex realm—and we are back to the reals for the rest of this section.

Example 10.2.3. *Write a program to implement Newton's Method.*

Solution: First let's write this algorithm in pseudo-code. For simplicity we assume

$$\frac{f(x_i)}{f'(x_i)}$$

is defined at each step. To help avoid the problem in the last example we add a counter n to make sure it does not loop forever in the while-loop.

10. Approximating Real Zeros: Numerical Methods

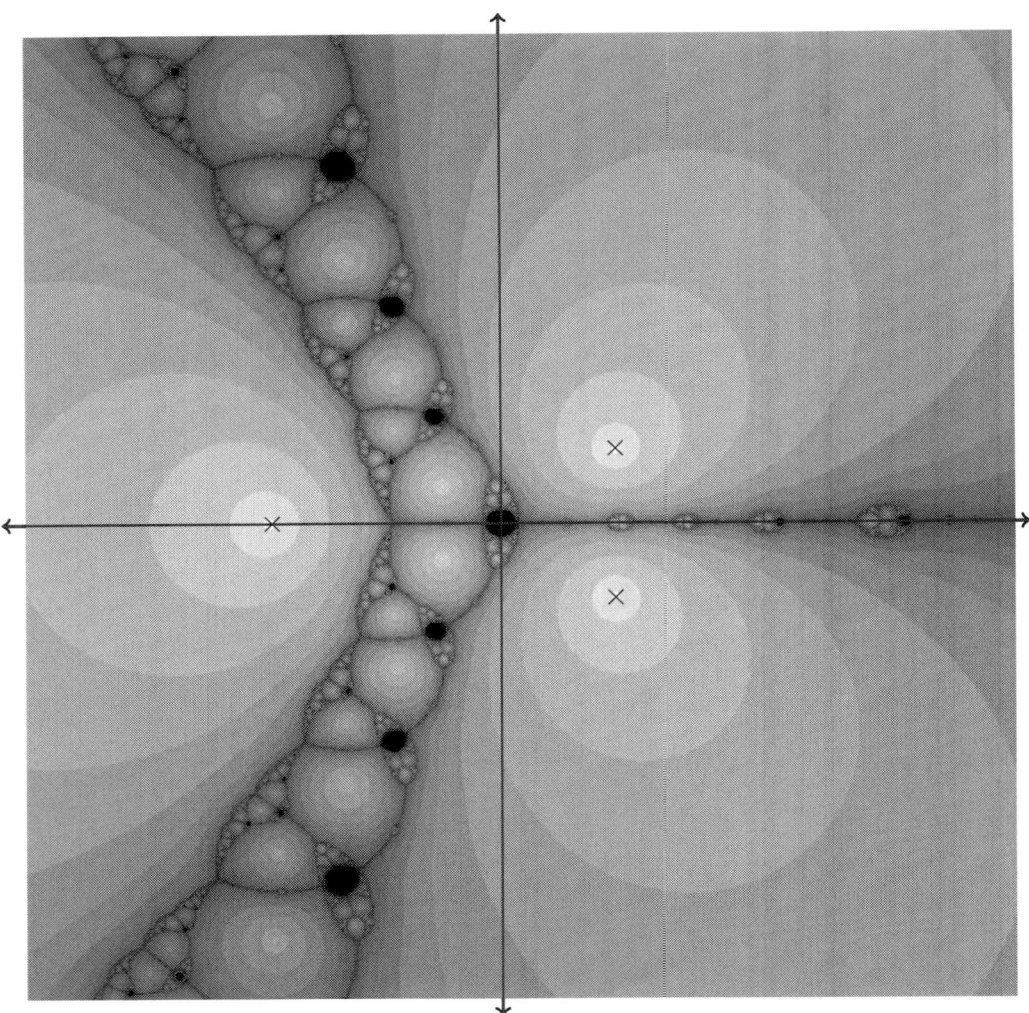

This graph shows the points $z = a + bi$ of the complex plane with $-4 \leq a \leq 4$ and $-4 \leq b \leq 4$. The three zeros of the polynomial $z^3 - 2z + 2$ are marked with the symbol ×. Each point z is shaded by how many iterations of Newton's Method it takes to get close to one of the zeros. The method converges quickly at the lightly shaded points and fails to converge at the black points. Each zero is surrounded by a "basin of attraction" (points for which the Newton's Method converge to that zero). The behavior where these regions meets is infinitely complex (the closer you look, the more complexity you see).

Figure 10.6: Newton Fractal for $z^3 - 2z + 2$

10.2. Faster But Not Sure: Newton-Raphson

Algorithm 10.2.1 Newton's Method

Require: $fprime$ is the derivative of f, initial value x_0 and tolerance ϵ

1: **procedure** NEWTON'S METHOD($f, fprime, x_0, \epsilon$)
2:     $n \leftarrow 0$
3:     **while** $|R - L| > \epsilon$ and $n < 100$ **do**
4:        $n \leftarrow n + 1$ ▷ Limit to 100 iterations.
5:        $x_1 \leftarrow x_0 - f(x_0)/fprime(x_0)$ ▷ The key step.
6:        **if** $|(x_1 - x_0)/x_1| < \epsilon$ **then**
7:            **return** x_1
8:        **end if**
9:        $x_0 \leftarrow x_1$ ▷ Just keep track of the last two values.
10:    **end while**
11:    **return** Failed to find a zero.   ▷ $n = 100$
12: **end procedure**

Now we will translate this to `Maple`, but you may use the language of your choice. We will assume $f(x)$ and $f'(x)$ are defined elsewhere so that the function may be easily changed. Also the program should accept the initial guess as input. In `Maple` we can write:

```
newton := proc(x,f,fp,epsilon)
    local x0, x1, n;
    x0 := x;
    n := 0;
    while n < 200 do
        x1 := x0-evalf(f(x0)/fp(x0));
        n := n+1;
        print(n, x1);
        if abs((x1-x0)/x1) < epsilon then
            return x1
        end if;
        x0 := x1
    end do;
    return "failed";
end proc;
```

■

Exercises 10.2:

1. Find a zero of the following functions and equations using Newton's method.

10. Approximating Real Zeros: Numerical Methods

   a) $x^2 - \cos x$

   b) $x + 1 + \sin x$

   c) $xe^x - 3$

   d) $x \log x = 15$

   e) $\sin x - \csc x = -1$

   f) $x^2 - 7$

   g) $x^3 - 56x + 1$

   h) $x^3 + 3x^2 - 4x - 12$

   i) $x^4 - 6x^3 + 12x^2 - 20x - 9$

   j) $x^6 + 6x^4 + 10x^2 + 3$

2. Show that the line tangent to $y = f(x)$ at the point $(x_i, f(x_i))$ intersects the x-axis at x_{i+1} as given in equation 10.1.

3. (*) Uspensky [30, pg. 179] gives the following variation of Newton's method which *always converges* regardless of the initial value:

$$x_{i+1} = x_i - \frac{f(x_i) f'(x_i)}{f'(x_i)^2 - 0.5 f(x_i) f''(x_i)}$$

   Write a program which approximates a zero of $f(x)$ using this method. The program should not require an initial guess and should calculate f, f' and f'' in subroutines.

4. (*) Another method for finding the zeros of f, called the **secant method**, is a variation of Newton's method. Rather than use the tangent line to $y = f(x)$ at $(x_i, f(x_i))$, use the secant line through $(x_i, f(x_i)), (x_{i-1}, f(x_{i-1}))$ (see problem 1). Find the formula for x_{i+1} in terms of x_i and x_{i-1}.

5. Repeat exercise 1 using the secant method or exercise 4.

6. How does the secant method of exercise 4 differ from problem Regula Falsi (exercise 8 of the previous section)?

10.3 A Gamble: Simple Iteration

To **iterate** means to do again and again. Newton's method is a form of iteration because we continually repeat the process by which we find x_{k+1} from x_k. More generally, to solve an equation $f(x) = 0$ by **simple iteration**, we solve the equation for x in terms of x; that is, we rewrite $f(x) = 0$ as $x = g(x)$ for some function $g(x)$. We then guess at a zero of $f(x)$ and call our guess x_0. If x_0 is zero then $x_0 = g(x_0)$, if not then maybe $x_1 = g(x_0)$ is closer to the zero, and $x_2 = g(x_1)$ even closer, ... The

10.3. A Gamble: Simple Iteration

numbers x_i are called the **iterates**. If the iterates x_i approach some limit x (approach a zero x), then we say that **the iterates converge** (or that the iteration converges). If the iterates x_i do not approach a limit we say that the **iterates diverge** (or that the iteration diverges).

Example 10.3.1. *Solve $x^3 - 6x + 2$ by simple iteration.*

Solution: We can solve for x (in terms of x) in several ways, among them we have the following.

$$x = \sqrt[3]{6x - 2} = g_1(x)$$

$$x = \sqrt{6 - \frac{2}{x}} = g_2(x)$$

$$x = \frac{6}{x} - \frac{2}{x^2} = g_3(x)$$

$$x = \frac{x^3 + 2}{6} = g_4(x)$$

We record the result of choosing 1 as the initial value (x_0) in Table 10.3. All four

x_i	g_1	g_2	g_3	g_4
x_0	1.00000000	1.00000000	1.00000000	1.00000000
x_1	1.58740091	1.99999989	4.00000000	0.50000000
x_2	1.95955467	2.23606801	1.37500000	0.35416666
x_3	2.13686442	2.25955128	3.30578511	0.34073743
x_4	2.21186328	2.26160765	1.63198750	0.33992672
x_5	2.24210763	2.26178551	2.92557413	0.33987975
x_{10}	2.26162147	2.26180219	2.04491638	0.33987689
x_{20}	2.26180219	2.26180219	2.22814412	0.33987689
x_{40}	2.26180219	2.26180219	2.26104101	0.33987689

Table 10.5: Iteration from $x_0 = 1$ (Example 10.3.1)

columns are clearly converging, the first three to the zero near 2.61802196502686, and the fourth to the zero near 0.3398768901824951. If we begin with the initial value $x_0 = 100$, the table looks like this: The first two columns again converge—but the latter two diverge. Once we have found any zero the other two can be found using the quadratic formula. ∎

After studying Example 10.3.1 the reader should be asking herself

- How can I tell if the iterations will converge or diverge?

10. Approximating Real Zeros: Numerical Methods

x_i	g_1	g_2	g_3	g_4
x_0	100.000000	100.000000	100.000000	100.000000
x_1	8.424943	2.445404	0.059800	166667.000000
x_2	3.648061	2.276431	-458.943412	$> 10^{14}$
x_3	2.709358	2.263057	-0.013830	
x_4	2.424752	2.261910	-12143.245121	
x_5	2.323793	2.618117	-0.000494	
x_{10}	2.262358	2.261802	$< -10^{23}$	

Table 10.6: Iteration from $x_0 = 100$ (Example 10.3.1)

- If the iterates converge, to which zero do they converge?
- If the iterates converge, how fast do they converge?

As a partial answer we offer the following: (for a more complete answer and the proof of this theorem, see almost any text on *numerical methods*, for example [29, chpt. 3]).

Theorem 10.3.1. *The iteration process $x_{k+1} = g(x_k)$ will converge to a zero of $x = g(x)$ if there is a real number q such that*

$$|g'(x)| \leq q < 1 \text{ (for } a \leq x \leq b) \quad \text{and} \quad a \leq x_0 \pm \frac{|g(x_0) - x_0|}{1 - q} \leq b.$$

Notice this theorem gives a sufficient condition but not a necessary condition.

The usual way to apply the method of simple iteration is to rearrange the equation to be solved (solving for x in terms of x) and then trying a few iterations to see if the process will converge. This method is very easy to use on a programable hand calculator because it involves no decision. For example, on a TI Calculator it can often be done by storing the initial value in x, then entering the formula and repeatedly pressing **ENTER**.

Exercises 10.3:

1. Find a zero of the following using simple iteration.
    a) $x^2 - \cos x = 0$
    b) $x + 1 + \sin x = 0$
    c) $xe^x - 3 = 0$
    d) $x \log x = 15$
    e) $x^3 - 56x + 1$

f) $x^3 + 3x^2 - 4x - 12$

g) $x^6 + 6x^4 + 10x^2 + 3$

h) $x^4 - 6x^3 + 12x^2 - 20x - 9$

2. The equation $x^3 - x = 0$ has the zeros 0 and 1. Which zero will we find if we iterate $x_{i+1} = x_i^3$? On what real interval will it converge?

3. (*) The equation $x^3 - x = 0$ has the zeros 0 and 1. Add $10x$ to each side and divide by 10. For what initial values x_0 is the iteration guaranteed to converge by Theorem 10.3.1?

4. (*) Show that Newton's method is just a special case of simple iteration. Is bisection also a special case of simple iteration? Why or why not?

5. (**) Use Theorem 10.3.1 of this section to find conditions under which Newton's method is guaranteed to converge.

6. (*) Give an example to show that Theorem 10.3.1 is sufficient but not necessary.

Numbers: Rational, Irrational, Algebraic and Transcendental

It is well known that the man who first made public the theory of irrationals perished in a shipwreck, in order that the inexpressible and unimaginable should ever remain veiled.

Proclus

Now as to what pertains to these Surd numbers (which, as it were by way of reproach and calumny, having no merit of their own, are also styled Irrational, Irregular, and Inexplicable) they are by many denied to be numbers properly speaking, and are wont to be banished from Arithmetic to another Science (which yet is no science) viz., algebra.

Isaac Barrow *Mathematical Lectures*, 1734.

Bless us, divine Number, thou who generated gods and men.

The Pythagoreans

Chapter 11

Numbers: Rational, Irrational, Algebraic and Transcendental

In this chapter we will study several important classes of real numbers: rational, irrational, algebraic and transcendental. Algebraic numbers, the zeros of polynomials with integer coefficients, are of obvious importance to this book! Some of these ideas are covered in number theory courses (or advanced algebra courses) in college, but we wanted a brief survey for teachers here. As usual, please feel free to scan, or even skip, proofs you find too difficult. There are just here to give you an understanding of the techniques involved, and to be blunt, because I find them beautiful.

11.1 Rational or Irrational?

A REAL NUMBER α IS **rational** if and only if it can be expressed as the ratio of two integers (for example $\frac{3}{2}$, $\frac{5}{1}$, $\frac{1}{-2}$); these are the numbers we have all dealt with since primary school. Those real numbers which are not rational are called **irrational**. We will show the irrationals include $\sqrt{2}$, e and π. Note that i is not real, so it is neither rational or irrational.

A major theme in this chapter will be approximating real numbers using rational numbers, but first we must make sure we can recognize rational numbers. This can be harder than you might guess. It took the very early mathematicians many years to discover irrational numbers even existed! We still do not know if many numbers such as $\pi + e$ and 2^e are rational or not (we guess not, but mathematics has little regard for guesses).

In this section we will discuss four methods of deciding whether a number α is rational or irrational.

11. Numbers: Rational, Irrational, ...

Method One: Using the Rational Zero Theorem

We will begin with a method we introduced in Section 5.1. It might be helpful to scan the examples there once again. Recall that if α is a zero of a polynomial $P(x)$, and α is rational, then we can find it using the Rational Zero Theorem 5.1.1. So if it turns out that $P(x)$ has no rational zeros, and α is real, then we will have shown α is irrational!

Example 11.1.1. *Show that each of $\sqrt{2}$, $\sqrt[3]{5}$ and $\sqrt{2} - \sqrt{3}$ are irrational.*

Solution: Notice that $\sqrt{2}$ is a zero of $x^2 - 2$; $\sqrt[3]{5}$ is a zero of $x^3 - 5$; and $\sqrt{2} - \sqrt{3}$ is a zero of $x^4 - 10x^2 + 1$. A quick check with the Rational Zero Theorem shows these three polynomials do not have rational zeros, so each of these numbers are irrational. (If you do not recall how to form a polynomial like the third, review example 8.4.1.)

∎

Note that $x^2 + 1$ also has no rational zeros, but as we remarked above its zeros $\pm i$ are not real and so are not rational or irrational. This is nothing strange, $\pm i$ are also neither positive nor negative (and neither even nor odd) for the same reason! There is a way to extend the notion of rational to the complex numbers; you could study it in a course on algebraic number theory. The notion of positive, however, cannot be extended to the complex numbers in any reasonable way.

Method Two: Using the Decimal Expansion

Look at the decimal expansions of the rational numbers in Table 11.1 and notice how the digits seem to repeat in blocks, but the expansions of the irrational numbers do not seem to do so. The number of digits in the (smallest) block of digits that are repeated (even if it is 0 that is repeated) is called the **period of α**. Decimal expansions that end with repeated blocks of zeros are called **terminating** (but note exercise 19).

The following theorem then gives us a way to prove numbers rational or irrational by knowing enough about their decimal expansions.

Theorem 11.1.1. *A real number α is rational if and only if the decimal expansion of α either terminates or is periodic (repeats) from some point onward.*

(The proof is found in exercises 15 and 16.)

Example 11.1.2. *Show that $X = 0.123456789101112131415\ldots$ (formed by letting the digits in X's decimal expansion be the integers $1, 2, 3, \ldots$) is irrational.*

11.1. Rational or Irrational?

Examples of Decimal Periods

α	decimal expansion	repeated digits	period of α	conclusion
1	$1.00000000000000000\overline{0}$	0	1	rational
$\dfrac{1}{2}$	$0.50000000000000000\overline{0}$	0	1	rational
$\dfrac{-1}{18000}$	$-0.00005555555555555\overline{5}$	5	1	rational
$\dfrac{2}{3}$	$0.66666666666666666\overline{6}$	6	1	rational
$\dfrac{3}{7}$	$0.428571428571\overline{428571}$	428571	6	rational
$\dfrac{59}{19}$	$3.\overline{105263157894736842}$	105263157894736842	18	rational
e	$2.781828182845904524\ldots$	—	—	irrational
$\sqrt{2}$	$1.414213562373095049\ldots$	—	—	irrational
π	$3.141592653589793238\ldots$	—	—	irrational
X	$0.123456789101112131\ldots$	—	—	irrational

Table 11.1: Example of Decimal Periods

Solution: By the choice of the digits of X, we know that it cannot eventually just repeat some block of digits infinitely because the integers never repeat, so X cannot be rational. ∎

Example 11.1.3. *Let $R = 123.456\overline{654321}$. Find integers a and b for which $a/b = R$.*

Solution: Since R has period 6, we multiply R by 10^6 and then subtract R:

$$\begin{aligned} 10^6 R &= 123456654.321654321654321654321654321\overline{654321} \\ R &= 123.456654321654321654321654321\overline{654321} \\ \hline 999999 R &= 123456530.865000000000000000000000000000 \end{aligned}$$

This approach, multiplying by ten raised to the period and subtracting, always yields a terminating decimal expansion. For our example, we must now multiply by 1000

223

11. NUMBERS: RATIONAL, IRRATIONAL, ...

to get an integer,
$$999999000R = 123456530865,$$

so we have found
$$R = \frac{123456530865}{999999000} = \frac{8230435391}{66666600}.$$

■

For further examples of this type, see [19, chapter 9].

Method Three: Using the Definition of the Number

Some numbers can be shown to be irrational directly from the way that they are defined.

Example 11.1.4. *Show that the number $\log_{10} 2$ is irrational.*

Solution: Suppose $\log_{10} 2$ is rational. Then we can write $\log_{10} 2 = a/b$ for some integers a and b. Since $\log_{10} 2 > 0$, we can choose a and b to be positive. By the definition of $\log_{10} 2$ we have $2 = 10^{a/b}$, so $2^b = 10^a$. This means 5 divides 10^a, but 5 can never divide 2^b (for any positive integer b), so we cannot have $2^b = 10^a$. This means that $\log_{10} 2$ is irrational. ■

Example 11.1.5. *Show that the number e (the base of the natural logarithms) is irrational.*

Solution: This takes substantially more work, partly because the definition of e is deeper than that of $\log_{10} 2$, but you can handle it! Here we give Fourier's proof.

Recall from calculus that
$$e = 1 + \frac{1}{1!} + \frac{1}{2!} + \frac{1}{3!} + \frac{1}{4!} + \cdots.$$

For proof by contradiction, we assume that $e = a/b$ with a, b positive integers. We know e is not an integer (and in fact $2 < e < 3$), so $b \neq 1$. Multiplying the equation above by $b!$ we get
$$b!e = b! + \frac{b!}{1} + \frac{b!}{2!} + \frac{b!}{3!} + \frac{b!}{4!} + \cdots.$$

By our assumption $e = a/b$, so $a = be$ is an integer and it follows that $b!e$ is also an integer. Notice that the partial sum
$$\alpha = b! + \frac{b!}{1} + \frac{b!}{2!} + \frac{b!}{3!} + \frac{b!}{4!} + \cdots + \frac{b!}{b!}$$

is a sum of integers, so α is an integer too. Together these mean $b!e - \alpha$ is an integer. However,

$$\begin{aligned} b!e - \alpha &= \frac{b!}{(b+1)!} + \frac{b!}{(b+2)!} + \frac{b!}{(b+3)!} + \frac{b!}{(b+4)!} + \cdots \\ &= \frac{1}{(b+1)} + \frac{1}{(b+1)(b+2)} + \frac{1}{(b+1)(b+2)(b+3)} + \cdots \\ &< \frac{1}{(b+1)} + \frac{1}{(b+1)^2} + \frac{1}{(b+1)^3} + \frac{1}{(b+1)^4} + \cdots \\ &= \frac{1}{b+1} \cdot \frac{1}{1 - \frac{1}{b+1}} = \frac{1}{b+1} \cdot \frac{b+1}{b} = \frac{1}{b}. \end{aligned}$$

We have now shown that $b!e - \alpha$ is an integer, but $0 < b!e - \alpha < \frac{1}{b} < 1$. There are no integers between 0 and 1, so this is clearly a contradiction! This means our original assumption that e is rational was incorrect, and it follows that e is irrational. ∎

Method Four: Using Approximation by Rationals

The Dutch astronomer Christiaan Huygens, in 1680, was working on a model of the planetary system, and from what he knew of the orbits of Earth and Saturn needed a gear ratio of 77708431/2640858. Constructing gears with this many teeth would be impossible, so he sought approximations involving smaller integers. Thus he began his study of continued fractions and eventually settled on the approximation 206/7. What interest us here is not such approximations in general, but that irrational numbers always have *infinitely* many good approximations but rationals do not.

But wait, what is a "good approximation" of a number α by rationals? We usually mean that the difference between α and its approximation p/q differ by at most a constant times $1/q^2$. The following theorem uses one version of this.

Theorem 11.1.2. *A real number α is irrational if and only if there are infinitely many rational numbers p/q (with p and q relatively prime integers) satisfying*

$$\left| \alpha - \frac{p}{q} \right| \leq \frac{1}{2q^2}.$$

Proof. Showing there are only finitely many such approximations for rational numbers α is easy (exercises 7 and 8), but we can show both directions at once with an interesting geometrical argument. To do this, we construct a "circle packing" by graphing the circles with centers $(p/q, 1/2q^2)$ and radii $1/2q^2$ for each pair p, q of relatively prime integers. This is illustrated in Figure 11.1.

11. Numbers: Rational, Irrational, ...

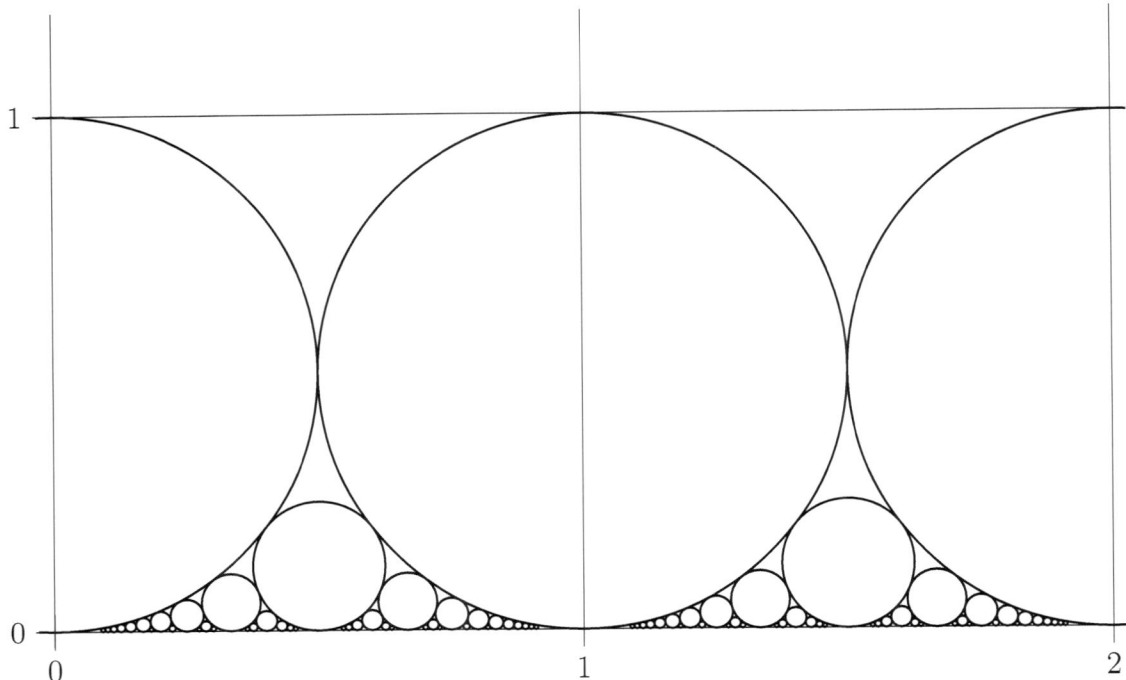

Figure 11.1: Circle Packing with Radius $\dfrac{1}{2q^2}$

Now visualize a vertical line $x = \alpha$. If α is rational, say $\alpha = p/q$ for relatively prime integers p and q, then the line only crosses finitely many of the circles and intersects the x-axis where the circle with center $(p/q, 1/2q^2)$ is tangent to the x-axis. Otherwise, if α is irrational, then no circle is tangent where the vertical line $x = \alpha$ intersects the x-axis, so this line intersects infinitely many of these packed circles. Why does intersecting such a circle matter? Because $x = \alpha$ intersects the circle with center p/q if and only if p/q approximates α to within $1/2q^2$. □

In Table 11.2 we list the first of π's infinite list of good rational approximations, and roughly half of them meet the criterion of Theorem 11.1.2. On the other hand, the rational $347/123$ has only the following five rational approximations which meet the criterion:

$$\frac{3}{1}, \frac{17}{6}, \frac{31}{11}, \frac{79}{28} \text{ and } \frac{347}{123}.$$

The conclusion of Theorem 11.1.2 is still true if $1/2q^2$ is replaced by the weaker condition $1/q^2$. All of the approximations of π in Table 11.2 meet this weaker criterion (as do infinitely many more); but again $347/123$ has only these ten:

$$\frac{2}{1}, \frac{3}{1}, \frac{15}{5}, \frac{17}{6}, \frac{31}{11}, \frac{48}{17}, \frac{79}{28}, \frac{110}{39}, \frac{268}{95} \text{ and } \frac{347}{123}.$$

11.1. Rational or Irrational?

p/q	approximation	error	q^2·error
3/1	3.0000000000000...	0.1415926535897...	0.1415...
22/7	3.1428571428571...	−0.0012644892673...	−0.0619...
333/106	3.1415094339622...	0.0000832196275...	0.9350...
355/113	3.1415929203539...	−0.0000002667641...	−0.0034...
103993/33102	3.1415926530119...	0.0000000005778...	0.6332...
104348/33215	3.1415926539214...	−0.0000000003316...	−0.3658...
208341/66317	3.1415926534674...	0.0000000001223...	0.5381...
312689/99532	3.1415926536189...	−0.0000000000291...	−0.2887...
833719/265381	3.1415926535810...	0.0000000000087...	0.6138...
1146408/364913	3.1415926535914...	−0.0000000000016...	−0.2144...
4272943/1360120	3.1415926535893...	0.0000000000004...	0.7474...

Table 11.2: Rational Approximations to π

Look up "continued fractions" to see how to find good rational approximations to any real number.

There is a bit of a cheat here. The reason you cannot approximate rationals more often is that they are their own best approximations (exercises 7 and 8). You just cannot do better than the rational itself!

Historical Note

The square root of two was shown to be irrational by the Pythagorean school about 500 BC. The discovery of irrationals was cataclysmic to the Pythagoreans, altering not only their method of proof, but also their theology. Curiously, Plato wrote in his *Theaetetus* that his teacher, Theodorus, extended this result by proving the irrationality of the square roots of 3, 5, 6, 7, 8, 10, ... up to 17, where he stopped. He obviously did not know the rational root theorem!

The numbers e and π were first shown to be irrational over 2000 years later by Lambert in 1761, and the number e^π was first shown to be irrational by Gelfond in 1929. Even today there are many numbers for which it is unknown if they are rational or irrational, for example, $\pi+e$, 2^e, $\ln \pi$, e^e, $e\pi$ and $\pi^{\sqrt{2}}$.

Exercises 11.1:

1. Prove that the (a) product; (b) sum, (c) difference and (d) quotient of two rational numbers are rational. (For the quotient, the denominator must be non-zero.)

2. Prove the following.

11. Numbers: Rational, Irrational, ...

> **Theorem 11.1.3.** *Let n, m be integers. Then $n^{1/m}$ is irrational unless n is the mth power of an integer k.*

Hint: See Example 11.1.1.

3. Write each of the following as a ratio of two integers.

    a) (a) $9283.37268372234234\overline{234}$
    b) (b) $-3.25377778888777788887\overline{7778888}$
    c) (c) $100000.00000000555\overline{5}$

4. Show that the following are irrational (a) $\log_2 3$, (b) $\log 3$, (c) $\log_3 7$, (d) $\log_2 \frac{5}{3}$.

5. Prove the following.

> **Theorem 11.1.4.** *Let n, m be positive integers, one of which has a prime factor that the other lacks. Then $\log_n m$ is irrational.*

Hint: See Example 11.1.4.

6. Show that if α is the zero of an irreducible polynomial of degree greater than one, then α is not rational.

7. Show that if a/b and p/q are rational numbers with

$$\left|\frac{a}{b} - \frac{p}{q}\right| < \frac{1}{bq},$$

then $\frac{a}{b} = \frac{p}{q}$. Hint: Just clear denominators...

8. Prove the following.

> **Lemma 11.1.5.** *If α is a rational number, and $\epsilon > 0$, then there are only finitely many rational numbers p/q for which $\left|\alpha - \frac{p}{q}\right| < \frac{1}{q^{1+\epsilon}}$.*

Hint: If $\alpha = a/b$, what must be true about q for $q^\epsilon > b$? Use that with the previous exercise. (Note that this shows that there are only finitely many solutions to the inequality in Theorem 11.1.2 for rational α.)

9. (*) Prove the following.

Lemma 11.1.6. *If α is a rational number, then there are infinitely many rational numbers p/q for which $\left|\alpha - \dfrac{p}{q}\right| \leq \dfrac{1}{q}$.*

Hint: Let $\alpha = a/b$ with $\gcd(a,b) = 1$. Look up the "Extended Euclidean Algorithm" and how it shows there are infinitely many solutions (p,q) to $aq + bp = 1$. Use this result and show the rationals p/q each satisfy the inequality. Does $\alpha = 0$ need special handling?

10. Show that (a) between every pair of rationals on the number line there is an irrational, and (b) between every pair of irrationals there is a rational. Hint: $a + (b-a)/\sqrt{2}$ and decimal expansions.[1]

11. (*) Prove that (a) $\sin 1$, and (b) $\cos 1$ ($1 = 1$ radian) are not rational. Hint: $\sin x$, $\cos x$ can be expanded as series expansion in the same way the e was. Look at how we proved e irrational.

12. Prove that $\sin 1°$ and $\cos 1°$ are not rational. Hint: If they were, explain why this would contradict Theorem 8.4.5.

13. (a) Find the periods of $1/p$ for the primes $p = 2, 3, 5, 7, 11, 13$, and 31. (b) Compare the periods of $1/p$ to $p-1$—what is the relationship? (Hint: divide.) (c) Does the relationship you found hold for all integers $n > 1$? What about $1/4$, $1/6$, $1/15$ and $1/21$?

14. (*) Give at least one plausible explanation why Theodorus stopped at 17. Hint: The answer to this question is not known, but several possible answers may be found in Heath's book *A History of Greek Mathematics* [20] and are repeated in [19, section 4.5].

15. (*) Prove that α is rational if the decimal expansion of α is periodic from some point onward. Hint: Let p be the period of α. Show that the decimal expansion of $(10^p x - x)$ is terminating (that is, it ends in a repeating string of zeros). Thus for some power m of 10, $10^m(10^p x - x)$ is an integer q, and $x = \dfrac{q}{10^m(10^p - 1)}$. This proves half of Theorem 11.1.1, the other half is the following problem.

16. (**) Prove that if α is rational, then the decimal expansion of α is periodic from some point onward. Hint: Let $\alpha = p/q$ for integers p, q with $q > 0$. Use Euler's Theorem below to show that q divides $10^g - 1$ for large enough integers g.

---

[1] Despite the impression this problem gives, there is a real sense in which there are far more irrational numbers than rational numbers. Look up Cantor's diagonalisation argument.

11. NUMBERS: RATIONAL, IRRATIONAL, ...

> **Theorem 11.1.7 (Euler's Theorem).** *Let a, m be any integers with no prime factor in common. Let $\phi(m)$ be the number of positive integers less than m which have no prime factor in common with m. Then m divides $a^{\phi(m)} - 1$.*

Instead of using this theorem, you can also use the Pigeonhole Principle.

17. (**) Prove then Theorem 11.1.1 is true regardless of the base of our number system.

18. Prove that $Y = 0.10100100010000100000100000100\ldots$ (using one zero, then two, then three...) is irrational.

19. (*) Prove that if α is a rational number, that the period of α is well defined. That is, show that if we have two different decimal expansions for α, they both have the same period. (Example: $0.5000000000000\ldots = 0.49999999999999\ldots$)

20. (*) Prove the following (to be used in the next two exercises).

> **Lemma 11.1.8.** *Let n be a positive integer. Define a function $f(x)$ and integers c_i ($n \leq i \leq 2n$) by*
>
> $$f(x) = \frac{x^n(1-x)^n}{n!} = \frac{c_n + c_{n+1}x + \ldots + c_{2n}x^{2n}}{n!}.$$
>
> *Show that each of $f(0)$, $f^{(m)}(0)$, $f(1)$, and $f^{(m)}(1)$ are integers; and that $0 < f(x) < 1/n!$ whenever $0 < x < 1$.*

Hint: Fill in the details: (a) $f(1-x) = f(x)$ so we need only prove the result for 0. (b) $f(0) = f^{(m)}(0) = 0$ if $m < n$ or $m > 2n$. (c) $f^{(m)}(0) = m!/(n! c_m)$, an integer. (d) $0 < x < 1$ implies $0 < f(x) < 1/n!$.

21. (**) Prove the following.

> **Theorem 11.1.9.** *e^y is irrational for every rational number $y \neq 0$.*

Hint: Fill in the details: Let $y = h/k$. If e^y is rational so are $e^{ky} = e^h$ and e^{-h}. So it is sufficient to prove this theorem for h a positive integer. Assume $e^h = a/b$ for proof by contradiction. Let $f(x)$ be as above and define $F(x)$ by

$$F(x) = h^{2n} f(x) - h^{2n-1} f'(x) + \ldots - h f^{(2n-1)}(x) + f^{(2n)}(x).$$

Notice the following is an integer, call it x.

$$\int_0^1 h^{2n+1} e^{hx} f(x) dx = be^{hx} F(x) \Big|_0^1 = aF(1) - bF(0).$$

Now using $0 < f(x) < 1/n!$, we see that the integral x above is less than $bh^{2n}e^h/n!$ so $0 < x < 1$ for n large enough. This contradicts x being an integer.

22. (**) Prove the following.

Theorem 11.1.10. *π and π^2 are irrational.*

Hint: Let $\pi^2 = \dfrac{a}{b}$ for proof by contradiction. Define $G(x)$ as follows:

$$G(x) = b^n \left(\pi^{2n} f(x) - \pi^{2n-2} f^{(2)}(x) + \pi^{2n-4} f^{(4)}(x) - \ldots + (-1)^n f^{(2n)}(x) \right).$$

Notice

$$\frac{d}{dx} \left(G'(x) \sin \pi x - \pi G(x) \cos \pi x \right) = b^n \pi^{2n+2} f(x) \sin \pi x = \pi^2 a^n f(x) \sin \pi x$$

and

$$\alpha = \pi \int_0^1 a^n \sin \pi x f(x) dx = \left. \frac{G'(x) \sin \pi x}{\pi} \right|_0^1 = G(0) + G(1).$$

So α is an integer, but $0 < \alpha < \pi a^n/n! < 1$ for large enough n, a contradiction. Finally, if π^2 is not rational, then neither is π.

11.2 Algebraic or Transcendental?

A number α (real or complex) is **algebraic** if it is the zero of a polynomial with rational coefficients, otherwise it is **transcendental**. For example, all rational numbers are algebraic (p/q is a zero of $qx - p$). Since most of this text involved solving such polynomials, we have already seen hundreds of examples of algebraic numbers!

If α is algebraic, then of all the non-zero polynomials with integer coefficients and with α as a zero, there is a least possible degree. This lowest degree is called the **degree of α**. See the examples in Table 11.2.

Recall that a rational polynomial (one with with rational coefficients) is said to be reducible (or reducible over the rationals) if and only if it is the product of two other polynomials with rational coefficients and lower degree; and that a polynomial is irreducible if it is not reducible. We leave the following simple consequence to the reader (exercise 11).

Theorem 11.2.1. *Let P be any irreducible polynomial with rational coefficients for which $P(\alpha) = 0$. Then the degree of α is the degree of P.*

11. NUMBERS: RATIONAL, IRRATIONAL, ...

α	polynomial of least degree with α as a zero	the degree of α
37	$x - 37$	1
$\sqrt{5}$	$x^2 - 5$	2
$\sqrt{2} + \sqrt{3}$	$x^4 - 10x^2 + 1$	4
$1 + i$	$x^2 - 2x + 2$	2
$(2 - 3\sqrt{5})/7$	$49x^2 - 28x - 41$	2
π	none	undefined

Table 11.3: Examples of Degrees of Algebraic Numbers

Again, the key way of spotting transcendental numbers is that algebraic numbers are more difficult to approximate with rationals. In fact Gelfond wrote:

> All methods of proof of the transcendence of a number in either the explicit or implicit form depend upon the fact that algebraic numbers cannot be very well approximated by rational fractions....

We will finish this section with the first of several theorems which illustrate this fact, starting with a theorem of Liouville from 1844.

Theorem 11.2.2. (Liouville's Theorem on Diophantine Approximation)
Let α be a real algebraic number of degree $n \geq 2$. There is a constant $c(\alpha) > 0$, depending only on α, such that

$$\left| \alpha - \frac{p}{q} \right| > \frac{c(\alpha)}{q^n}$$

for all integers p and q.

Liouville's theorem makes it possible to easily list an infinite number of transcendental numbers (see exercise 7). We'll begin with one of the easiest examples.

Example 11.2.1. *Show the following number is transcendental.*

$$\alpha = 2^{-1} + 2^{-2} + 2^{-6} + 2^{-24} + \cdots + 2^{-n!} + \cdots$$

Solution: Suppose α is algebraic, so has some finite degree n. We will show this causes a contradiction. Define $c(\alpha)$ as in Liouville's Theorem, and let P_k and Q_k, for any integer $k > n$, be defined as follows.

$$Q_k = 2^{k!}$$
$$P_k = 2^{k!} \left(2^{-1} + 2^{-2} + 2^{-6} + 2^{-24} + \cdots + 2^{-n!} \right)$$

By Liouville's Theorem

$$\left|\alpha - \frac{P_k}{Q_k}\right| = 2^{-(k+1)!} + 2^{-(k+2)!} + 2^{-(k+3)!} + \cdots > (???)\frac{c(\alpha)}{Q_k^n}.$$

Multiplying by $Q_k^n = 2^{n \cdot k!}$, we see that $c(\alpha)$ is less than

$$2^{k!(n-(k+1))} + 2^{k!(n-(k+1)(k+2))} + 2^{k!(n-(k+1)(k+2)(k+3))} + \cdots$$

$$\leq 2^{-k!} + 2^{-k!-1} + 2^{-k!-2} + \cdots = 2 \cdot 2^{-k!}.$$

for every integer $k > n$. Choosing k to be large we may make this upper bound on $c(\alpha)$ smaller than any positive number, so $c(\alpha)$ is zero. This contradicts Liouville's theorem, so α is not algebraic. ∎

Proof of Liouville's Theorem. The theorem requires that α be an algebraic number of degree $n > 1$, so (by exercise 10), α is a zero of of a polynomial with integer coefficients

$$P(x) = a_n x^n + a_{n-1} x^{n-1} + \cdots + a_1 x + a_0$$

where $a_n \neq 0$. The degree of α is n, so α cannot be a zero of a polynomial of any lower degree, which means this polynomial is irreducible and hence has no rational zeros (we will use this below).

Choose any upper bound M of $|P'(x)|$ on the interval $\alpha-1 \leq x \leq \alpha+1$. (Here $P'(x)$ is the derivative of $P(x)$.) We will show that any constant less than $D = \min(1, 1/M)$ will work as the desired constant $c(\alpha)$ (which we will call simply C for the remainder of this proof).

Let p and q be any integers with $q > 0$. If $\left|\alpha - \frac{p}{q}\right| \geq 1$, then we are done because

$$\left|\alpha - \frac{p}{q}\right| \geq 1 \geq D > C \geq \frac{C}{q^n}$$

since q is a positive integer and $C < D \leq 1$. In the rest of the proof we will assume $\left|\alpha - \frac{p}{q}\right| < 1$.

As mentioned above, $P(x)$ has no rational zeros, so $P(p/q) \neq 0$ and it follows

$$\left|P\left(\frac{p}{q}\right)\right| = \frac{|a_n p^n + a_{n-1} p^{n-1} q + \cdots + a_1 p q^{n-1} + a_0 q^n|}{q^n} \geq \frac{1}{q^n}$$

(because the numerator is a non-zero integer). The Mean Value Theorem (Theorem 9.1.3) now shows

$$\left|P\left(\frac{p}{q}\right)\right| = \left|P\left(\frac{p}{q}\right) - P(\alpha)\right| = \left|\frac{p}{q} - \alpha\right| |P'(\beta)|$$

11. Numbers: Rational, Irrational, ...

for some number β between p/q and α. Together these give

$$\frac{1}{q^n} \leq \left| P\left(\frac{p}{q}\right) \right| \leq \left| \alpha - \frac{p}{q} \right| M,$$

so

$$\left| \alpha - \frac{p}{q} \right| \geq \frac{1}{M} \frac{1}{q^n} \geq \frac{D}{q^n} > \frac{C}{q^n}.$$

This completes the proof. □

Exercises 11.2:

1. Let $\alpha = \sqrt{2}$. Find both of the error measures $|\alpha - p/q|$ and $q^2|\alpha - p/q|$ for the following rational approximations p/q of α: 1/1, 3/2, 7/5, 17/12, 41/29, 99/70, 239/169, 577/408. (See Table 11.2 for an example). What pattern do you see?

2. Let $\gamma = (1+\sqrt{5})/2$. Find both of the error measures $|\gamma - p/q|$ and $q^2|\gamma - p/q|$ for the following rational approximations p/q of γ: 2/1, 3/2, 5/3, 8/5, 13/8, 21/13, 34/21, 55/34. What pattern do you see? [By the way, do you recognize the numbers in these fractions? In a measurable sense γ is the most difficult irrational to approximate.]

3. What are the degrees (if any) of the following?

    a) 27
    b) 0
    c) e
    d) $\log_2 8$
    e) $\sqrt{2}$
    f) $\sqrt[4]{84}$
    g) $\sqrt[3]{5}$
    h) $\sqrt{2} - \sqrt{3}$
    i) $\sqrt[3]{2}\sqrt[5]{3}$
    j) $\log_{10} 2$
    k) $3^{7/4}$
    l) $1 + \sqrt{2}$
    m) i
    n) $2 + i$
    o) $\sqrt{3} - \sqrt{-6}$
    p) i^{15}
    q) $2^{3/17}$

4. Show that the degree of $\sqrt{2} + \sqrt{3} + \sqrt{6}$ is 4.

5. Show that the degree of $\sqrt{2} + \sqrt{3} + \sqrt{5}$ is 8.

6. Show that Liouville's constant is transcendental:

    $$\frac{1}{10} + \frac{1}{10^2} + \frac{1}{10^6} + \frac{1}{10^{24}} + \cdots + \frac{1}{10^{n!}} + \cdots = 0.110001000000000000000000100\ldots.$$

    Hint: This was his example for his theorem.

7. (*) Let A and B be any fixed positive integers. Let $a_1, a_2, a_3, \ldots$ be any sequence of positive integers for which $a_n \leq A$ for $n = 1, 2, 3, \ldots$. Show that the number

    $$\gamma = \frac{a_1}{B} + \frac{a_2}{B^2} + \frac{a_3}{B^{3!}} + \frac{a_1}{B^{4!}} + \cdots + \frac{a_n}{B^{n!}} + \cdots$$

    is transcendental. Hint: Example 11.2.1 is the case $A = 1$ (so $a_n = 1$ for all n) and $B = 2$. Exercise 6 was the case $A = 1$ and $B = 10$.

11.3. Measuring Irrationality

8. (**) A **Liouville number** is an irrational number x such that for every positive integer n there are integers p and q with $q > 1$ and such that
$$0 < \left| x - \frac{p}{q} \right| < \frac{1}{q^n}.$$
Prove there are infinitely many Liouville numbers. Hint: Use the previous problems? If you know what 'uncountable' means, you can show there are uncountably many at the same time.

9. Prove that all Liouville numbers (as defined in the previous problem) are transcendental. Hint: The namesake's theorem?

10. In the definition of algebraic number, show that "with rational coefficients" can be changed to "with integer coefficients." (This is the way the definition is stated in many books.)

11. (*) Prove Theorem 11.2.1. Hint: Use the following outline.

    a) The degree of α is less than or equal to the degree of P.
    b) Let Q be any polynomial with α as a zero. Then α is a zero of $\gcd(P, Q)$.
    c) Show $\gcd(P, Q) = P$.
    d) The degree of α is the degree of P.

12. Suppose α and β are algebraic numbers and q is any rational number. Show the following are also algebraic numbers.

    a) $1/\alpha$ (if $\alpha \neq 0$)   Hint: Theorem 6.2.3
    b) $q\alpha$ Hint: Theorem 6.2.1
    c*) $\alpha + \beta$ Hint: Use Exercise B.1.5
    d*) $\alpha\beta$ Hint: Use Exercise B.1.6

11.3 Measuring Irrationality

We want to end this chapter, and this book, with two final results. We will start by defining a measure of how well numbers can be approximated by rationals, and then end by stating a powerful test for the transcendence of a special form of number.

The **irrationality measure** β of a real number α is the smallest real number such that
$$\left| \alpha - \frac{p}{q} \right| > \frac{1}{q^{\beta + \epsilon}}$$
holds for any $\epsilon > 0$ and all integers p and q with q sufficiently large. This is same as saying β is the smallest real number such that for every $\epsilon > 0$ there is a constant

$A(\epsilon)$ such that
$$\left| \alpha - \frac{p}{q} \right| < \frac{A(\epsilon)}{q^\beta}$$
has only finitely many solutions in rational numbers p/q.

The irrationality measure is defined to be infinite if no such real number β exists.

Example 11.3.1. *Show that the irrationality measure of a rational number is one.*

Solution: Let α be a rational number. Given any $\epsilon > 0$, by Lemma 11.1.5 there are only finitely many solutions to $\left| \alpha - \frac{p}{q} \right| < \frac{1}{q^{1+\epsilon}}$, thus $\left| \alpha - \frac{p}{q} \right| > \frac{1}{q^{1+\epsilon}}$ for q sufficiently large. This shows the irrationality measure of α is at most one.

By Lemma 11.1.6, $\left| \alpha - \frac{p}{q} \right| < \frac{1}{q}$ infinitely often, so the irrationality measure of α cannot be less than one. Together, these show the irrationality measure is one. ∎

Example 11.3.2. *Show that the irrationality measure of an irrational number is at least two.*

Solution: Suppose, for proof by contradiction, that α is irrational and has irrationality measure $\beta < 2$. Choose $\epsilon > 0$ so that $\beta + \epsilon < 2$, then for q large enough, $q^{\beta+\epsilon-2} < 2$. This means $q^{\beta+\epsilon} < 2q^2$, so
$$\left| \alpha - \frac{p}{q} \right| < \frac{1}{2q^2} < \frac{1}{q^{\beta+\epsilon}}.$$

But Theorem 11.1.2 says this inequality has infinitely many solutions, a contradiction, showing the irrationality measure of α is at least two. ∎

Liouville's Theorem 4.2.6 shows that an algebraic number with degree n has an irrational measures at most n. But it is possible very greatly improve on that. For years mathematicians worked to reduce the bounds on these measures; this effort finally culminated in the following very deep theorem.

Theorem 11.3.1 (Roth, 1955). *Let α be a real algebraic number. Then for any $\epsilon > 0$, the inequality*
$$\left| \alpha - \frac{p}{q} \right| < \frac{1}{q^{2+\epsilon}}$$
has only finitely many solutions in relatively prime integers p and q.

So all irrational algebraic numbers have irrationality measure exactly two.

11.3. Measuring Irrationality

We have seen that rationals have irrationality measure one, the other algebraic numbers two; so wouldn't it be great if we could say x is transcendental if its irrationality measure is greater than two? Sadly this is not true. The irrationality measure of the transcendental number e is also exactly two (see [10]). The best we can say is that if α is transcendental, then its irrationality measure is at least two. Some transcendentals have far higher irrationality measure as shown in the next example.

Example 11.3.3. *Show that Liouville numbers have infinite irrationality measures.*

Solution: Suppose α is a Liouville number, so by the definition of Liouville number we can define a sequence of rationals p_n/q_n for which

$$\left| \alpha - \frac{p_n}{q_n} \right| < \frac{1}{q_n^n}. \tag{11.1}$$

There must be infinitely many different rationals in this sequence (because if any one rational p/q worked for infinitely many n's, then $\alpha = p/q$). This means that inequality 11.1 has infinitely many solutions for every integer n (namely p_m/q_m for all $m \geq n$).

So given any real number β, we can choose an integer $n > \beta$ and we have an infinite number of solutions to

$$\left| \alpha - \frac{p_n}{q_n} \right| < \frac{1}{q_n^n} < \frac{1}{q^\beta}.$$

So β cannot be the irrationality measure—this means the measure is infinity. ∎

Lindemann-Weierstrass Theorem

Though the proof is well beyond the level of this book, we wish to end by presenting Baker's reformulation of the Lindemann-Weierstrass Theorem. From this theorem it is easy to see that π and e are transcendental.

Theorem 11.3.2 (Lindemann-Weierstrass Theorem). *If $a_1, \ldots, a_n$ are nonzero algebraic numbers, and $\alpha_1, \ldots, \alpha_n$ are distinct algebraic numbers, then*

$$a_1 e^{\alpha_1} + \cdots + a_n e^{\alpha_n} \neq 0.$$

As powerful as that theorem is, it is unknown if any of the following numbers are rational or irrational (let alone if they are transcendental!):

$$\pi + e, \quad \pi - e, \quad \pi e, \quad \frac{\pi}{e}, \quad e^e, \quad e^{e^e}, \quad \pi^e, \quad \pi^{\sqrt{2}} \quad \text{or} \quad \ln \pi.$$

11. NUMBERS: RATIONAL, IRRATIONAL, ...

Exercises 11.3:

1. Use Theorem 11.3.2 to show π is transcendental. Hint: Recall $e^{\pi i} + 1 = 0$.

2. Use Theorem 11.3.2 to show e is transcendental. (Notice this exercise and the previous one are far simpler than the corresponding proofs that e and π are irrational in the previous section—but here we are cheating, all of the complexity is hidden in the Lindemann-Weierstrass Theorem.)

3. Using the fact that e is transcendental, prove that $\ln r$ is irrational for all rational numbers r other than one. Hint: If $\ln r = \dfrac{p}{q}$, then $e^{p/q} = r$. So e is the zero of the following *polynomial* $x^p = r^q$ with integer coefficients.

4. Suppose α and β are transcendental numbers and q is any non-zero rational number. (a) Show by example that $\alpha + \beta$ and $\alpha\beta$ need not be transcendental. (b) Prove that $1/\alpha$ and $q\alpha$ are transcendental.

5. (***) Show that we cannot use an exponent larger than 2 in Theorem 11.1.2 by filling in the details in the following.

   Let $\epsilon > 0$. If we create circles as in the proof of Theorem 11.1.2, but with the smaller radii $1/2q^{2+\epsilon}$, then gaps appear between neighboring circles because they are no longer tangent (see Figure 11.2).

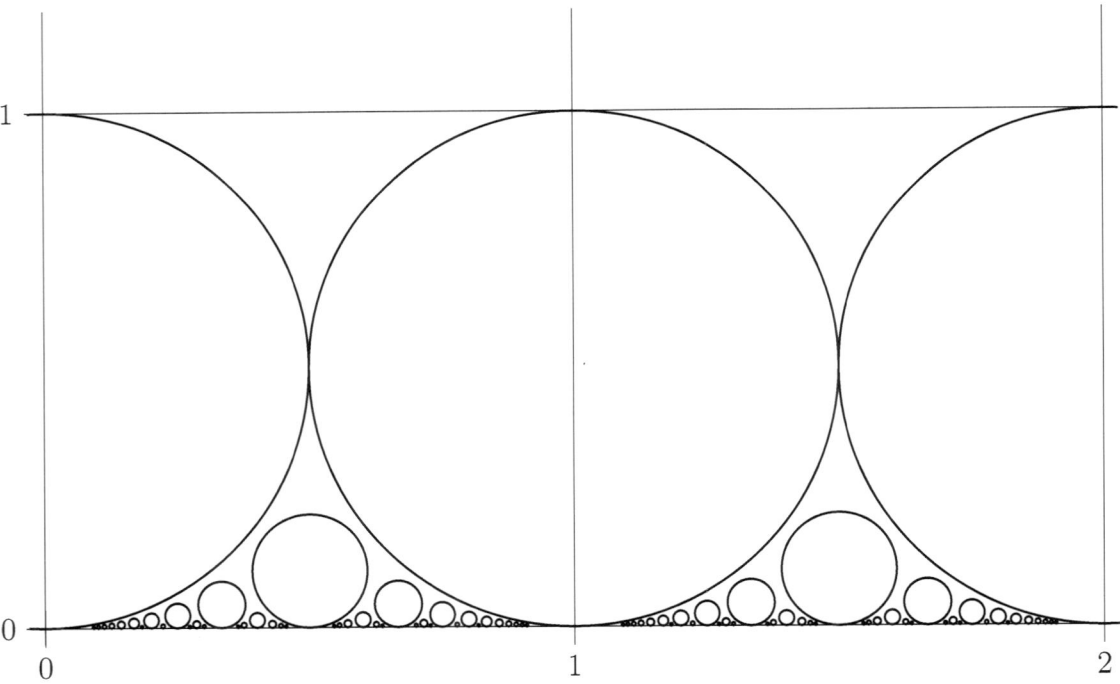

Figure 11.2: Circle Packing: Radius $\dfrac{1}{2q^{2+\epsilon}}$

11.3. Measuring Irrationality

Choose any two consecutive integers and consider just the circles that represent pairs p/q between them. The circle for p/q contributes a width of $2/2q^{2+\epsilon}$, so the total width of all such circles is less than

$$\sum_{q=1}^{\infty} \frac{1}{q^{1+\epsilon}} < 1 + \int_1^{\infty} \frac{dq}{q^{1+\epsilon}} = 1 + \frac{1}{\epsilon}.$$

This is finite, so almost every vertical line intersects only finitely many circles.

Appendices

Appendix A

Delayed Proofs

A.1 The Fundamental Theorem of Algebra

Theorem A.1.1 (The Fundamental Theorem of Algebra). *Every polynomial of degree greater than zero has at least one zero.*

In 1746 d'Alembert first stated and attempted to prove this theorem. The first rigorous proof was given by Gauss in his 1799 thesis. Since then dozens of proofs have been published and continue to be published. Those readers who have studied complex analysis will recall that the Fundamental Theorem of Algebra is a trivial consequence of Liouville's Theorem (problem 1).

In this section we give an intuitive (informal) sketch of Gauss' fourth and final proof. For an *almost* complete version see Uspensky's *Theory of Equations* [30].

To begin the proof let the polynomial be written

$$P(x) = x^n + a_1 x^{n-1} + a_2 x^{n-2} + \cdots + a_n. \tag{A.1}$$

We then write x and the coefficients of P in their polar forms $x = r(\cos\theta + i\sin\theta)$,

$$a_k = A_k(\cos\theta_k + i\sin\theta_k) \qquad \text{for} \quad k = 1, 2, \ldots, n.$$

Using De Moivre's Theorem 2.5.1, and the angle addition formulas, we find

$$a_k x^{n-k} = A_k \left(\cos\theta_k + i\sin\theta_k\right) r^{n-k} \left(\cos((n-k)\theta) + i\sin((n-k)\theta)\right)$$
$$= A_k r^{n-k} \left(\cos((n-k)\theta + \theta_k) + i\sin((n-k)\theta + \theta_k)\right).$$

So now we may separate $P(x)$ into real and imaginary components

$$P(x) = T(r,\theta) + iU(r,\theta) \tag{A.2}$$

A. Delayed Proofs

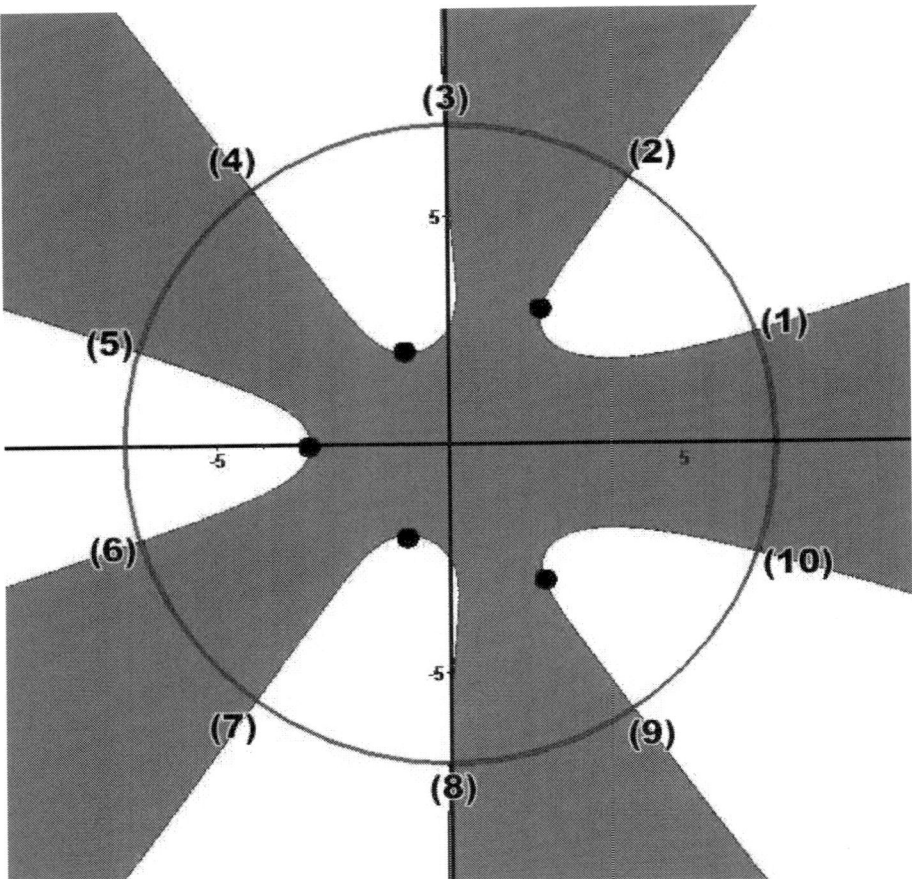

Figure A.1: Where $T(r, \theta)$ is Positive for an Example Quintic

where T and U are the following real valued functions.

$$T(r,\theta) = r^n \cos(n\theta) + A_1 r^{n-1} \cos((n-1)\theta + \theta_1) + \cdots + A_n \cos \theta_n$$
$$U(r,\theta) = r^n \sin(n\theta) + A_1 r^{n-1} \sin((n-1)\theta + \theta_1) + \cdots + A_n \sin \theta_n$$

To prove the theorem, we must find a value of x, or equivalently, values of r, θ, for which both U and T are simultaneously zero. This will show that $P(x) = T(x) + iU(x) = 0$. We do this by studying the behavior of U and T for large values of r, that is, values of x which are far from the origin. Notice as r gets very large the first terms of T and U become the dominant terms, that is, the terms with greatest absolute value. In fact if r is large enough, the sign of $T(r, \theta)$ is the sign of $r^n \cos n\theta$, and the sign of $U(r, \theta)$ is the sign of $r^n \sin n\theta$.

To make this more precise we could prove the following (problem 2).

A.1. The Fundamental Theorem of Algebra

Lemma A.1.2. *Let $P(x)$ be as in equation A.1. There is a real number R such that if $r > R$, then*

$$r^n - \sqrt{2}(a_1 r^{n-1} + a_2 r^{n-2} + \cdots + a_n) > 0.$$

The proof of this lemma is straight forward; in fact $R = 2 + \sqrt{2}C$ will do if C is the largest of the absolute values of $1, A_1, A_2, \ldots, A_n$.

Notice the expression in Lemma A.1.2 is not quite T or U. The reason the trigonometric functions have disappeared and the square root of 2 appears is buried in the first use of this lemma, which is proving the following.

Lemma A.1.3. *Let $T(r, \theta)$ be as in equation A.2. The circumference of a circle of radius $r > R$ consists of $2n$ arcs on which T is alternately positive and negative.*

What this lemma is saying may be made clearer by viewing Figure A.1. This figure shows where T is positive and negative for an example polynomial (a monic quintic, the black circles represent the five zeros -3, $2 \pm 3i$ and $-1 \pm 2i$). The points on the plane which are shaded are the complex numbers at which T is negative, the unshaded points are where T is positive, and the curves between the shaded and unshaded regions are where T is zero. We see that the circle is divided into $2n$ (in this case ten) arcs where T is alternately positive and negative.

Idea of the proof of Lemma A.1.3. Consider the $4n$ points with amplitude $\frac{(2i+1)\pi}{4n}$ ($i = 1, 2, \ldots, 4n - 1$). These points are evenly spaced about the circle being considered. To find the sign of T at these points we need only find the sign of its leading coefficient (thanks to Lemma A.1.2)

$$r^n \cos\left(\frac{n(2i+1)\pi}{4n}\right) = r^n \cos\left(\frac{(2i+1)\pi}{4}\right) = \pm\frac{r^n}{\sqrt{2}}.$$

where the sign is positive if $i = 1, 4, 5, 8, 9, \ldots, 4n - 3$ and negative if $i = 2, 3, 6, 7, 10, 11, \ldots, 4n - 1$. Thus T changes sign between the first and second point, between the third and fourth point, between $\ldots$, that is, T changes sign $2n$ times. □

From Lemma A.1.3 we may conclude that there are $2n$ points on this circle where T is zero. Call these points $(1), (2), \ldots, (2n)$. In the same way that we prove Lemma A.1.3 we may also show the following.

Lemma A.1.4. *With the notation above, U is positive at at the points $(1), (3), \ldots, (2n-1)$ and negative at $(2), (4), \ldots, (2n)$.*

A. Delayed Proofs

Proof of the Theorem A.1.1: Lemma A.1.3 shows that for $r > R$, the circle of circumference r may be divided into $2n$ regions where T is alternately positive and negative. As r increases and the circle is increased, these arcs sweep out regions as shown in Figure A.1. Let's say that the shaded areas (the areas where T is positive) are seas, and the unshaded areas are lands. If we begin at an odd numbered point, say at (1), and walk along the shore always keeping the land to our right, we eventually must cross the circle again and at an even numbered point, in this case (2). For example, in Figure A.1 we may have walked from (1) to (2), from (3) to (4), from (5) to (6), from (7) to (8) or from (9) to (10). Now on our walk along the shore, T is always zero and by Lemma A.1.4, U changes from negative to positive—so somewhere on our walk U must have been zero. This is the desired point x at which both T and U are simultaneously zero. □

Exercises A.1:

1. (*) A function is called **analytic** if it has a derivative at every point for which it is defined. Prove the Fundamental Theorem of Algebra using the following result from the course Complex Variables.

    Theorem A.1.5 (Liouville's Theorem). *A function which is both analytic and bounded on the whole complex plane must be a constant.*

    Hint: Assume $P(x)$ has no zeros and consider $1/P(x)$.

2. Prove Lemma A.1.2.

3. Use Theorem A.1.1 to prove Lemma 4.2.2

4. Look up (in another book) and state the Fundamental Theorem of Arithmetic. In this text we have also listed two "fundamental theorems" (A.1.1 and 6.4.1). What do these three theorems have in common?

A.2 Proof of Sturm's Theorem

In this section we prove Sturm's Theorem 9.5.1.

Proof of Sturm's Theorem. First let's recall the notation in the theorem. Here $(a, b]$ is an interval of real numbers, $P(x)$ is a non-zero polynomial and

$$P_0 = P, \quad P_1 = P', \quad P_2, \quad P_3, \quad \ldots, \quad P_k,$$

its Sturm sequence. By the way Sturm sequences are defined we have polynomials $Q_i(x)$ (the quotients from the Euclidian Algorithm) such that

$$P_{i-1}(x) = Q_i(x)P_i(x) - P_{i+1}(x). \tag{A.3}$$

The key to this proof is to note that only way for $\text{Var}(P, t)$ to change value as t increases from a to b, is for one of the terms $P_i(t)$ to change sign, and this means that $P_i(t) = 0$ for some i and some $t \in (a, b)$. By Equation A.3, if two consecutive terms in the sequence are zero, say $P_i(t)=P_{i+1}(t)=0$, then every term vanishes at t. In particular $P(t)=P'(t)=0$, and t is a multiple zero of $P(x)$ (see section 4.4).

For the rest of this proof suppose $P_i(t) = 0$ for some $0 \leq i \leq k$. We will consider two cases: first that t is not a multiple zero of P, then that it is.

First suppose that $x = t$ is not a multiple zero of $P(x)$, but $P_i(t) = 0$ for some $0 \leq i \leq k$. If $i > 0$, then by Equation A.3, $P_{i-1}(t)$ and $P_{i+1}(t)$ have opposite signs. Hence for all x near t, there is only one change in sign in the triple $P_{i-1}(x), P_i(x), P_{i+1}(x)$, so no change in $\text{Var}(P, x)$ near t. If instead $i = 0$, then $P'(t) \neq 0$ (because t is a simple zero), so $P(x)$ and $P'(x)$ have different signs for $x < t$ and the same signs for $x > t$ (see Figure A.2).

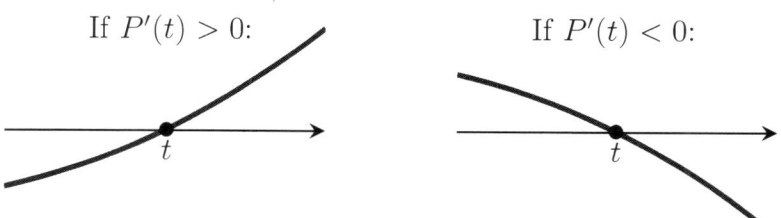

Figure A.2: The Graph of $P(x)$ Near the Simple Zero t

This shows $\text{Var}(P, x)$ decreases by one as x passes each simple zero t.

This leaves us the case t is a multiple zero of $P(x)$. Let the multiplicity of t be $m+1$, then $(x - t)^m$ divides $P'(x)$ and also every term of the Sturm sequence by equation A.3. Consider instead the sequence:

$$P_0/(x-t)^m, \quad P_1/(x-t)^m, \quad P_2/(x-t)^m, \quad P_3/(x-t)^m, \quad \ldots, \quad P_k/(x-t)^m.$$

This sequence also satisfies Equation A.3 for $i \geq 1$. For $x \neq t$, the number of variations in sign in this sequence is the same as in the Sturm sequence of P (because we have multiplied every term by the same non-zero real). And as before, zeros of any term other than the first will not alter the number of changes of sign, so we need to just consider when the first term is zero.

If we write $P(x) = (x - t)^{m+1}Q(x)$ for some polynomial $Q(x)$, then

$$P'(x) = (m + 1)(x - t)^m Q(x) + (x - t)^{m+1} Q'(x).$$

A. Delayed Proofs

The fist two terms in the sequence are then

$$(x-t)Q(x) \quad \text{and} \quad (m+1)Q(x) + (x-t)Q'(x).$$

If $Q(t) > 0$, then the second term is positive near t and the sign of the first changes from negative to positive. If instead $Q(t) < 0$, then the second term is negative near t and the sign of the first changes from positive to negative. Either way, the number of variations in sign decreases by one as x passes through the multiple zero t of P. This completes the proof of Sturm's Theorem. □

A.3 Bundan-Fourier and Descartes' Rule

Given a polynomial of degree n, define δP to be the sequence of derivatives

$$P(x),\ P'(x),\ P^{(2)}(x),\ P^{(3)}(x),\ \ldots\ P^{(n)}(x)$$

and let $\text{Var}(\delta P, x)$ be the number of changes of sign in this sequence when each term is evaluated at x. As usual, $\text{Var}(\delta P, \pm\infty)$ can be determined by the leading coefficients. In particular, $\text{Var}(\delta P, \infty) = 0$? because the leading coefficients of every term have the same sign.

> **Theorem A.3.1 (Bundan-Fourier Theorem).** *Suppose that $P(x)$ is a polynomial with real coefficients, (a, b) is an interval of real numbers, and $z(P, (a, b])$ is the number of zeros of P in $(a, b]$ (counting multiplicity). Then*
>
> $$\text{Var}(\delta P, a) - \text{Var}(\delta P, b) \geq Z(P, (a, b])$$
>
> *and the difference is even.*

The proof of Descartes' Rule of Sign gives a nice simple example of this theorem.

Proof of Theorem 9.2.1. Let $P(x)$ be a polynomial with real coefficients. Since we find $\text{Var}(\delta P, \infty)$ by checking leading coefficients of

$$P(x),\ P'(x),\ P^{(2)}(x),\ P^{(3)}(x),\ \ldots\ P^{(n)}(x);$$

and these all have the same sign, $\text{Var}(\delta P, \infty) = 0$. This means that the number of positive zeros is at most $\text{Var}(\delta P, 0)$ (and if less, then less by an even number). But $\text{Var}(\delta P, 0)$ is determined by the sequence of signs in the constant terms of δP, that is, the sequence of the signs of the coefficients of $P(x)$.

Example A.3.1. *Count the real zeros of $P(x) = x^4 - 2x^3 + 2x^2 - 2x + 1$.*

A.3. Bundan-Fourier and Descartes' Rule

Solution: By Descartes' Rule, there are 4, 2 or no positive zeros and no negative zeros. The sequence δP is

$$x^4 - 2x^3 + 2x^2 - 2x + 1, \quad 4x^3 - 6x^2 + 4x - 2, \quad 12x^2 - 12x + 4, \quad 24x - 12, \quad 24.$$

At $x=2$ this is $5, 14, 28, 36, 24$. This presents no changes, like at ∞, so there are no zeros in $(2, \infty)$. At $x=0$ this sequence is $1, -2, 4, -12, 24$, presenting four changes of sign. So the Bundan-Fourier theorem tells us there are 4, 2 or no zeros in the interval $(0,2)$.

The Sturm sequence for $P(x)$ is

$$P(x), \quad P'(x), \quad -\tfrac{1}{4}x^2 + x - \tfrac{3}{4}, \quad -32x + 32.$$

At $x=0$ this is $1, -2, -3/4, 32$ (two changes), and at $x=2$ this is $5, 14, 1/4, -32$, (one change). So there is exactly one zero in the interval $(0, 2)$.

It looks like there is a conflict here. One theorem says there is one zero; and the other says "4, 2 or 0." But there is no conflict: Burdan-Fourier's theorem counts multiplicity and Sturm's theorem does not, so there is one zero and its multiplicity is two or four. In fact $P(x) = (x-1)^2(x^2+1)$. ∎

Proof of the Bundan-Fourier Theorem. As x increases across the given interval $(a, b]$, $\text{Var}(\delta P, x)$ only can change at the zeros $x=t$ of some derivative $P^{(i)}(x)$ $(0 \leq i < k)$. (Here $i=0$ give P as the zeroth derivative.) Let t be such a zero and choose $\epsilon > 0$ small enough that no other term of the sequence δP has a zero in $0 < |x-t| \leq \epsilon$. Let $n \geq 0$ be the multiplicity of t as a zero of P. Will show the following two claims.

- $\text{Var}(\delta P, t) = \text{Var}(\delta P, t + \epsilon)$, and

- $\text{Var}(\delta P, t - \epsilon) \geq \text{Var}(\delta P, t) + n$ (and the difference is even).

The proof of the theorem follows easily from these two statements, because as t increases from a to b, $Z(P, (a, t]) + \text{Var}(\delta P, t)$ only changes at the zeros of P and its derivatives, and at each of them it changes by the order of t as a zero of $P(x)$ plus an even number.

We will prove these two claims above by induction on the degree of $P(x)$. For the remainder of this proof t is a zero of some derivative $P^{(i)}(x)$ $(i \geq 0)$ and $\epsilon > 0$ is small enough that $(t - \epsilon, t + \epsilon)$ contains no other zeros of and of the $P^{(i)}(x)$ $(i \geq 0)$.

If the degree of P is one, then: P' is a non-zero constant; the sign of P' does not change; and t is a zero of P $(i = 0)$. Now either $P' > 0$ and P changes from negative to positive at t, or $P' < 0$ and P changes from positive to negative.

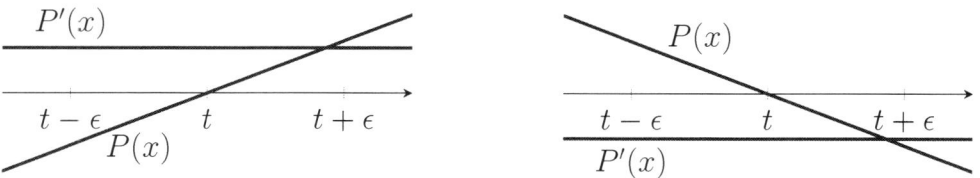

A. Delayed Proofs

Either way $\mathrm{Var}(\delta P, t - \epsilon) = 1$ and $\mathrm{Var}(\delta P, t) = \mathrm{Var}(\delta P, t + \epsilon) = 0$. This proves the two claims in this case.

For proof by induction we suppose that the results have been proven for polynomials of degree m, and let $P(x)$ have degree $m + 1$. Case I. Suppose $P(t) = 0$ and let $n \geq 0$ be the multiplicity of $x = t$ as a zero of $P(x)$. So $x = t$ is a zero of P' of order $n1$. By our induction hypothesis we have the following.

- $\mathrm{Var}(\delta P', t) = \mathrm{Var}(\delta P', t + \epsilon)$, and
- $\mathrm{Var}(\delta P', t - \epsilon) \geq \mathrm{Var}(\delta P', t) + m - 1$ (and the difference is even).

By the Mean Value Theorem 9.1.3, P and P' have opposite signs to the left of t and the same signs on the right (see exercise 2). This shows

- $\mathrm{Var}(\delta P, t) = \mathrm{Var}(\delta P, t + \epsilon)$, and
- $\mathrm{Var}(\delta P, t - \epsilon) = \mathrm{Var}(\delta P', t - \epsilon) + 1 \geq \mathrm{Var}(\delta P, t) + m$

and the difference is even. This completes the proof if $P(t) = 0$.

Case II: Now suppose that $P(t) \neq 0$ and let $n \geq 0$ be the multiplicity of $x = t$ as a zero of $P'(x)$. So $P'(t) = P''(t) = \cdots = P^{(n)} = 0$; but $P^{(n+1)}(t) \neq 0$. Multiplying P by -1 if necessary[1] we may assume $P^{(n+1)}(t) > 0$ so it follows that $P^{(i)}(t + \epsilon) > 0$ for $1 \leq i \leq n$. Finally, by our induction hypothesis applied to P', we again have the following.

- $\mathrm{Var}(\delta P', t) = \mathrm{Var}(\delta P', t + \epsilon)$, and
- $\mathrm{Var}(\delta P', t - \epsilon) \geq \mathrm{Var}(\delta P', t) + n$ (and the difference is even).

Suppose first that n is even. Note that $P'(t \pm \epsilon) > 0$, so for each y in $\{t - \epsilon, t, t + \epsilon\}$ the first non-zero term in

$$P'(y),\ P''(y),\ P^{(3)}(y),\ P^{(4)}(y),\ \ldots$$

is positive.

Examples of Case II with n even:

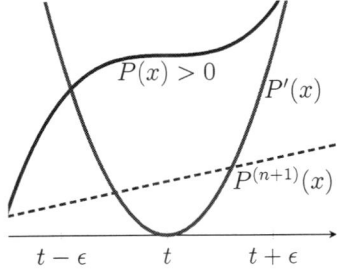 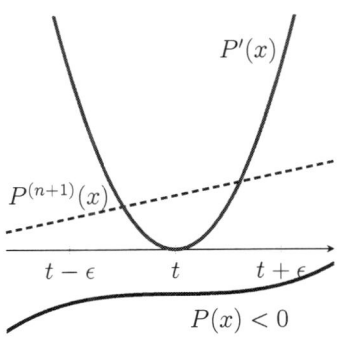

---

[1]Multiplying $P(x)$ by -1 does not alter the number of sign changes in the sequence because it changes the sign of every term.

When $P(y)$ is positive, this gives $\text{Var}(\delta P, y) = \text{Var}(\delta P', y)$ and when $P(y)$ is negative, it gives $\text{Var}(\delta P, y) = \text{Var}(\delta P', y) + 1$. This proves the two claims because it shows $\text{Var}(\delta P, t) = \text{Var}(\delta P', t + \epsilon)$ and

$$\text{Var}(\delta P, t - \epsilon) - \text{Var}(\delta P, t) = \text{Var}(\delta P', t - \epsilon) - \text{Var}(\delta P', t)$$

which is n plus an even number, hence is even.

Finally suppose n is odd. So $P'(t - \epsilon) < 0$ and $P'(t + \epsilon) > 0$ so the first non-zero term in the subsequence

$$P'(y),\ P''(y),\ P^{(3)}(y),\ P^{(4)}(y),\ \ldots$$

is $-$ at $Y = t - \epsilon$ and $+$ at $y = t$ and $y = t + \epsilon$ respectively.

Examples of Case II with n odd:

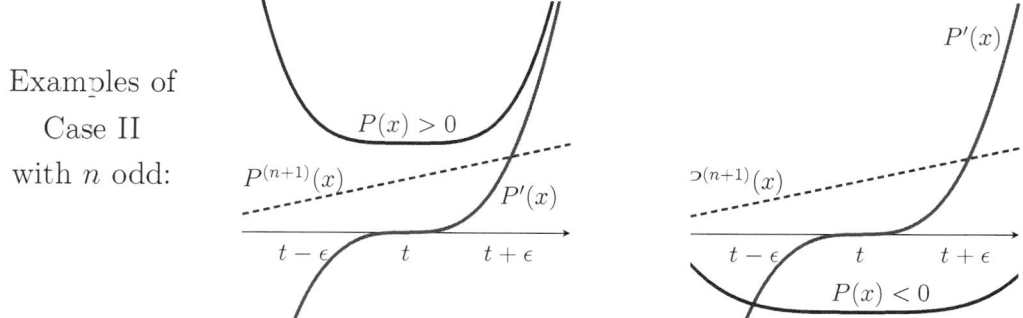

When $P(t)$ is positive, this gives $\text{Var}(\delta P, t - \epsilon) = \text{Var}(\delta P', t - \epsilon) + 1$; and also $\text{Var}(\delta P, y) = \text{Var}(\delta P', y)$ at both $y = t$ and $y = t + \epsilon$. When $P(t)$ is negative, it gives $\text{Var}(\delta P, t - \epsilon) = \text{Var}(\delta P', t - \epsilon)$; but $\text{Var}(\delta P, y) = \text{Var}(\delta P', y) + 1$ at $y = t$ and $y = t + \epsilon$. Either way $\text{Var}(\delta P, t) = \text{Var}(\delta P', t + \epsilon)$ and

$$\text{Var}(\delta P, t - \epsilon) - \text{Var}(\delta P, t) = \text{Var}(\delta P', t - \epsilon) - \text{Var}(\delta P', t) \pm 1.$$

This last difference is an odd number (n) plus or minus one, so must be even. This completes the proof. □

Exercises A.3:

1. For each of the following polynomials P, find limits on the number of positive and negative zeros. Then use Bundan-Fourier Theorem to find intervals each of which contains just one zero of P.

    a) $x^3 - 3x^2 - 11$

    b) $x^3 - 3x + 1$

    c) $x^4 - 5x^3 + 5x^2 + 4x - 2$

    d) $3x^5 + x^3 - 2x^2 - 2x + 1$

A. Delayed Proofs

e) $2x^6 + 4x^5 - 14x^2 + 6x + 1$

f) $x^8 + x^3 - 8x^2 + 2x + 2$

2. Suppose that $P(x)$ is a polynomial, that P as a zero at t (of any multiplicity ≥ 1) but neither P nor P' have other zeros in the interval $(t - \epsilon, t + \epsilon)$. Show that $P(x)$ and $P'(x)$ have opposite signs if $t - \epsilon < x < t$ and the same sign if $t < x < t + \epsilon$. Hint: By the Mean Value Theorem 9.1.3

$$P'(c) = \frac{P(t + \epsilon) - P(t)}{\epsilon} = \frac{P(t + \epsilon)}{\epsilon}$$

for some c between t and $t + \epsilon$.

Appendix B

Useful Results

B.1 The Resultant of Two Polynomials

The resultant of two polynomials

$$p(x) = a_0 x^n + a_1 x^{n-1} + \cdots + a_{n-1} x + a_n$$

and

$$q(x) = b_0 x^m + b_1 x^{m-1} + \cdots + b_{m-1} x + b_m$$

(with $a_0 b_0 \neq 0$) is defined to be

$$\operatorname{Res}(p,q) = a_0^m b_0^n \prod_{i,j} (\alpha_i - \beta_j) \tag{B.1}$$

where the α_i are the zeros of $p(x)$ and the β_j are those of $q(x)$. If you do not recognize this notation, it just means create the mn differences of the form:

$$\text{(a zero of } p(x)) - \text{(a zero of } q(x)),$$

and then multiply them all together. From this definition we have the following theorem.

Theorem B.1.1. $\operatorname{Res}(p,q)$ *is zero if and only if $p(x)$ and $q(x)$ have a common zero, that is, if* $\gcd(p(x), q(x)) \neq 1$.

All we need to make this useful is an easy way to find the resultant!

The resultant $\operatorname{Res}(p,q)$ is a clearly a symmetric function of the zeros of the two polynomials, so it can be found as a function of the coefficients (Theorem 6.4.1). One

B. USEFUL RESULTS

way to do that is by using the determinant of the **Sylvester matrix** (called the Sylvester determinant) which is defined as below.

$$\begin{vmatrix} a_0 & a_1 & \cdots & a_{n-1} & a_n & 0 & \cdots & 0 \\ 0 & a_0 & a_1 & \cdots & a_{n-1} & a_n & \cdots & 0 \\ \vdots & \vdots & \vdots & \vdots & \vdots & \vdots & \vdots & \vdots \\ 0 & \cdots & 0 & a_0 & a_1 & \cdots & a_{n-1} & a_n \\ b_0 & b_1 & \cdots & b_{m-1} & b_m & 0 & \cdots & 0 \\ 0 & b_0 & b_1 & \cdots & b_{m-1} & b_m & \cdots & 0 \\ \vdots & \vdots & \vdots & \vdots & \vdots & \vdots & \vdots & \vdots \\ 0 & \cdots & 0 & b_0 & b_1 & \cdots & b_{m-1} & b_m \end{vmatrix} \begin{matrix} \left.\begin{matrix} \\ \\ \\ \\ \end{matrix}\right\} m \text{ rows} \\ \left.\begin{matrix} \\ \\ \\ \\ \end{matrix}\right\} n \text{ rows} \end{matrix}$$

This expression for $\mathrm{Res}(p, q)$ looks very elaborate, but it is actually quite simple: each row of this $(m+n)$ by $(m+n)$ matrix is the coefficients of one of the two polynomials padded with zeros. Let's look at a couple examples and see how the coefficients of the polynomials form the rows of the Sylvester determinant.

Example B.1.1. *Let $p(x) = ax^2 + bx + c$ and $q(x) = 2ax + b$, then we have*

$$\mathrm{Res}(p, q) = \begin{vmatrix} a & b & c \\ 2a & b & 0 \\ 0 & 2a & b \end{vmatrix} = -a(b^2 - 4ac).$$

Example B.1.2. *Let $p(x) = x^3 + bx + c$ and $q(x) = 3x^2 + b$, then we have*

$$\mathrm{Res}(p, q) = \begin{vmatrix} 1 & 0 & b & c & 0 \\ 0 & 1 & 0 & b & c \\ 3 & 0 & b & 0 & 0 \\ 0 & 3 & 0 & b & 0 \\ 0 & 0 & 3 & 0 & b \end{vmatrix} = 4b^3 + 27c^2.$$

Notice how close these are to the discriminants of the quadratic and cubic respectively.

Note: If you are not sure how to calculate these determinants and the example you are addressing is concrete (no variables), don't worry, just use your calculators!

Exercises B.1:

1. First find the resultants of the following pairs of polynomials to see if they are relatively prime, then check your answer using a different method.

    a) $x^2 - 2x + 1$ and $x^2 - 5x + 6$

    b) $x^2 - x$ and $x^2 - 3x + 2$

    c) $x^3 + x^2 + x + 1$ and $x^2 + x + 1$

B.1. The Resultant of Two Polynomials

2. Explain why $\mathrm{Res}(p(x), p'(x))$ gives us a value which is zero if and only if $p(x)$ has multiple zeros (a 'discriminant'). Here $p'(x)$ is the derivative of $p(x)$ from calculus.

3. The **discriminant** D of a polynomial is defined by
$$(-1)^{n(n-1)/2} a_0 D = \mathrm{Res}(p, p').$$
Use the Sylvester determinant to show the discriminate for a general cubic polynomial is
$$D = 18 a_0 a_1 a_2 a_3 - 4 a_1^3 a_3 + a_1^2 a_2^2 - 4 a_0 a_2^3 - 27 a_0^2 a_3^2.$$

4. Find the discriminant (see the previous problem) for
$$p(x) = x^4 + bx^2 + cx + d.$$
Hint: Enter the Sylvester matrix into a program such as Maple or Mathematica, and then use that program to evaluate the determinant.

5. Let $p(x)$ and $q(x)$ be polynomials of degrees n and m respectively. Show that the resultant $\mathrm{Res}(p(x), q(t-x))$ is a polynomial of degree mn in the variable t, with the mn zeros $\alpha_i + \beta_j$. Hint: Equation B.1.

6. Let $p(x)$ and $q(x)$ be polynomials of degrees n and m respectively. Show that the resultant $\mathrm{Res}(p(x), x^n q(t/x))$ is a polynomial of degree mn in the variable t, with the mn zeros $\alpha_i \beta_j$. Hint: Equation B.1.

7. Explain why the previous problems show that the sum and product of two algebraic numbers is algebraic.

8. Show that $\mathrm{Res}(p, q_1 q_2) = \mathrm{Res}(p, q_1) \mathrm{Res}(p, q_2)$ for polynomials p, q_1 and q_2. Hint: Equation B.1.

9. Let $p(x)$ and $q(x)$ be polynomials of degrees n and m respectively. Show that $\mathrm{Res}(p, q) = a_0^m \prod_i q(\alpha_i) = (-1)^{mn} = b_0^n \prod_j q(\beta_j)$.

10. (*) Show there are polynomials $r(x)$ and $s(x)$ of degrees not more than $m-1$ and $n-1$ respectively such that
$$r(x) p(x) + s(x) q(x) = R(p, q).$$
Hint: Solve $A\vec{v} = \vec{b}$ with Cramers rule, where A is the Sylvester matrix, $\vec{v}$ is the column vector $(x^{n+m-1}, x^{n+m-2}, \ldots, x, 1)$, and $\vec{b}$ is the column vector
$$(x^{m-1} p(x), \ldots, xp(x), p(x), x^{n-1} q(x), \ldots, xq(x), q(x)).$$

11. (**) Recall that you can find the rank of a matrix, the number of linearly independent rows, by row reducing it (e.g., by using `rref` on your TI calculator). Show that we can use the rank r of the Sylvester matrix to find the degree of the greatest common divisor of p and q because $\deg(\gcd(p,q)) = m + n - r$.

B. Useful Results

B.2 Division with Detached Coefficients

In Chapter Three we used synthetic division to divide polynomials by linear terms. This was done by "detaching" the coefficients and letting the column each number was in determine what it represents. For example, to divide

$$p(x) = 5x^6 - 10x^5 + 10x^4 - 18x^2 - 4x + 8$$

by $x - 2$ (or equivalently, to evaluate $p(2)$) we can proceed as follows.

$$
\begin{array}{r|rrrrrrr}
2 & 5 & -10 & 10 & 0 & -18 & -4 & 8 \\
  &   & 10 & 0 & 20 & 40 & 44 & 80 \\
\hline
  & 5 & 0   & 10 & 20 & 22 & 40 & \multicolumn{1}{|r}{88}
\end{array}
$$

This method is so quick and easy that it is taught in many secondary school classes and is often used by mathematicians when they want to do a quick evaluation. For many problems it is quicker than using software.

In this section we present an old way of extending this **method of detached coefficients** to polynomials with arbitrary degrees. I must admit however, that it is often easier to use technology than the method presented here. Still, for completeness it is left here as the answer to the question: can we do "synthetic division" with non-linear polynomials?

This method is best illustrated by examples.

Example B.2.1. *Divide $x^3 + 8x^2 + 3$ by $x^2 + x + 1$.*

Solution: We first use the 'old' method of long division:

$$
\begin{array}{r}
x + 7 \\
x^2 + x + 1 \overline{\smash{)} x^3 + 8x^2 + 3} \\
\underline{-x^3 - x^2 - x } \\
7x^2 - x + 3 \\
\underline{-7x^2 - 7x - 7} \\
-8x - 4
\end{array}
$$

The method of detached coefficients involves exactly the same work except for the following traditional changes:

1. only the coefficients are written down (without the "x"s and exponents),
2. the divisor is written to the left of the dividend (not to the right), and
3. the quotient is written under the divisor (instead of above the dividend).

So our work on this problem is arranged as follows.

B.2. Division with Detached Coefficients

$$
\begin{array}{rrr|rrr}
1 & 8 & 0 & 3 & 1 & 1 & 1 \\
1 & 1 & 1 & & & 1 & 7 \\
\hline
 & 7 & -1 & 3 & & & \\
 & 7 & 7 & 7 & & & \\
\hline
 & & -8 & -4 & & &
\end{array}
$$

This shows the quotient is $x+7$ and the remainder is $-8x-4$. (Compare this method carefully with the long division above.) ∎

Example B.2.2. *Divide $x^4 - 2x^3 + 3x^2 - 2x + 1$ by $x^2 - 3x + 1$.*

Solution: Again, the process of detached coefficients is very similar to long division. We begin both methods by writing down the dividend and quotient (here we show both methods side by side).

$$
\begin{array}{rrrrr|rrr}
1 & -2 & 3 & -2 & 1 & 1 & -3 & 1
\end{array}
\qquad x^2 - 3x + 1 \,\overline{\smash{)}\, x^4 - 2x^3 + 3x^2 - 2x + 1}
$$

We see that the first coefficient of the quotient is one, so write one down (or x^2), then multiply and subtract 1 times the divisor from the dividend.

$$
\begin{array}{rrrrr|rrr}
1 & -2 & 3 & -2 & 1 & 1 & -3 & 1 \\
1 & -3 & 1 & & & 1 & & \\
\hline
 & 1 & 2 & -2 & & & &
\end{array}
\qquad
\begin{array}{r}
x^2 \\
x^2 - 3x + 1 \,\overline{\smash{)}\, x^4 - 2x^3 + 3x^2 - 2x + 1} \\
-x^4 + 3x^3 - x^2 \\
\hline
x^3 + 2x^2 - 2x
\end{array}
$$

We see that the next coefficient of the quotient is also one and continue as follows.

$$
\begin{array}{rrrrr|rrr}
1 & -2 & 3 & -2 & 1 & 1 & -3 & 1 \\
1 & -3 & 1 & & & 1 & 1 & \\
\hline
 & 1 & 2 & -2 & & & & \\
 & 1 & -3 & 1 & & & & \\
\hline
 & & 5 & -3 & 1 & & &
\end{array}
\qquad
\begin{array}{r}
x^2 + x \\
x^2 - 3x + 1 \,\overline{\smash{)}\, x^4 - 2x^3 + 3x^2 - 2x + 1} \\
-x^4 + 3x^3 - x^2 \\
\hline
x^3 + 2x^2 - 2x \\
-x^3 + 3x^2 - x \\
\hline
5x^2 - 3x + 1
\end{array}
$$

Now we see the last quotient is 5 and finish as follows.

$$
\begin{array}{rrrrr|rrr}
1 & -2 & 3 & -2 & 1 & 1 & -3 & 1 \\
1 & -3 & 1 & & & 1 & 1 & 5 \\
\hline
 & 1 & 2 & -2 & & & & \\
 & 1 & -3 & 1 & & & & \\
\hline
 & & 5 & -3 & 1 & & & \\
 & & 5 & -15 & 5 & & & \\
\hline
 & & & 12 & -4 & & &
\end{array}
\qquad
\begin{array}{r}
x^2 + x + 5 \\
x^2 - 3x + 1 \,\overline{\smash{)}\, x^4 - 2x^3 + 3x^2 - 2x + 1} \\
-x^4 + 3x^3 - x^2 \\
\hline
x^3 + 2x^2 - 2x \\
-x^3 + 3x^2 - x \\
\hline
5x^2 - 3x + 1 \\
-5x^2 + 15x - 5 \\
\hline
12x - 4
\end{array}
$$

B. Useful Results

Thus the quotient is $x^2 + x + 5$ and the remainder is $12x - 4$. ∎

Exercises B.2:

1. Use the method of detached coefficients to find the quotient and remainder when the first polynomial is divided by the second. Then graph both the ratio of the two polynomials and the quotient on the same graph for $-10 < x < 10$. What do you notice about the two graphs? How could you use this to draw better graphs?

   a) $x^4 + 3x^2 + 6x - 4, x^2 - 2x + 2$
   b) $x^4 - 3x^3 + 3x^2 + 6x - 4, x^2 + 2x + 1$
   c) $x^5 - 32, x^2 + 2x$

   This approach to graphing (with simpler polynomials) is useful even for many graphs in secondary education courses.

2. Use the method of detached coefficients to find the quotient and remainder when the first polynomial is divided by the second.

   a) $x^4 - 4x^3 + 6x^2 - 4x + 1, x^2 - 2x + 1$
   b) $x^4 - 4x^3 + 6x^2 - 4x + 1, x^2 + 2x + 1$
   c) $x^5 - 16x, x^3 - 4x$
   d) $x^4 - x^2 + 1, x^2 + x + 1$
   e) $x^5 + x^4 + x^3 + x^2 + x + 1, 3x^3 + 3$
   f) $3x^5 + 2x^4 + 4x - 6, x^2 - 1$
   g) $2x^5 + 3x^4 + 5x^3 + 8x^2 + 13x + 21, x^2 + x + 1$
   h) $x^6 + 2x^5 + 3x^4 + 4x^3 + 8x^2 + 12x, x^3 + 4$
   i) $3x^4 + 2x^2 - 6x - 3, 4x^3 + 2x - 1$
   j) $5x^4 + 3x^3 + x^2 + 4x + 6, x^5 + 2$
   k) $x^6 + 3x^5 - x^4 - 2x^3 - 24x^2 + 6, x^2 + x - 6$
   l) $5x^5 + 4x^4 + 3x^3 + 2x^2 + x + 1, 3x^2 + 2x + 1$
   m) $x^4 + 3x^3 + x^2 - 3x - 2, x^3 + 2x^2 - x - 2$
   n) $ax^5 + bx^4 + cx^3 + cx^2 + bx + a, x^2 + 1$

3. There are many other situations in which we "detach coefficients." Explain what the methods given in this section have in common with using matrices to solve systems of linear equations.

Appendix C

Selected Answers

Exercises 2.1 ▲ Page 10

1. (a) If $z \in \mathbb{R}$, then $z = a + (0)i = a = a - (0)i = \bar{z}$. If $z = \bar{z}$, we get $a + bi = a - bi \Rightarrow bi = -bi \Rightarrow b = 0$, so $z = a + (0)i \Rightarrow z \in \mathbb{R}$.
(b) Assume z is purely imaginary, then $z = bi = -(-b)i = -\bar{z}$. Now suppose $z = -\bar{z}$. We get that $a + bi = -(a - bi) \Rightarrow a + bi = -a + bi \Rightarrow a = -a \Rightarrow a = 0$, then z is purely imaginary.

2. (a) $7 + 4i$, (g) $4 + 3i$, (m) $\cos^2\theta - i^2\sin^2\theta = 1$

3. (b) Here $m \in \mathbb{Z}$ $(-i)^n = \begin{cases} -i & n = 4m+1 \\ -1 & n = 4m+2 \\ i & n = 4m+3 \\ 1 & n = 4m \end{cases}$

(e) $\left(\frac{1+i}{\sqrt{2}}\right)^3 = \begin{cases} \frac{1+i}{\sqrt{2}} & n = 8m+1 \\ i & n = 8m+2 \\ \frac{-1+i}{\sqrt{2}} & n = 8m+3 \\ -1 & n = 8m+4 \\ \frac{-1-i}{\sqrt{2}} & n = 8m+5 \\ -i & n = 8m+6 \\ \frac{1-i}{\sqrt{2}} & n = 8m+7 \\ 1 & n = 8m \end{cases}$

4. (a) $x = -\frac{1}{2}, y = -6$

Exercises 2.2 ▲ Page 13

1. (a)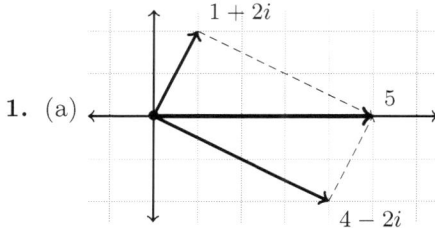

2. (a) $\sqrt{2554}$ (c) 1 (e) 1

3. $|z_1||z_2| = \sqrt{a_1^2 + b_1^2}\sqrt{a_2^2 + b_2^2}$
$= \sqrt{(a_1^2 + b_1^2)(a_2^2 + b_2^2)}$
$= \sqrt{z_1 z_2}$
$= |z_1 z_2|$

7. $|z_1| = |z_1 - z_2 + z_2| \leq |z_1 - z_2| + |z_2|$
$\Rightarrow |z_1| \leq |z_1 - z_2| + |z_2|$
$\Rightarrow |z_1| - |z_2| \leq |z_1 - z_2|$

9. (a) $a \pm bi$ are reflections of each other across the x-axis. (c) $\pm z$ are reflections through the origin. (e) The set of all points outside of the unit circle $|z| = 1$.

Exercises 2.3 ▲ Page 19

1. $r_1(\cos\theta_1 + i\sin\theta_1)r_2(\cos\theta_2 + i\sin\theta_2)$
$= r_1 r_2(\cos\theta_1 \cos\theta_2 + i\cos\theta_1 \sin\theta_2 + i\cos\theta_2 \sin\theta_1 - \sin\theta_1 \sin\theta_2)$
$= r_1 r_2(\cos(\theta_1 + \theta_2) + i\sin(\theta_1 + \theta_2))$

3. If $z_1 = a_1 + b_1 i$ and $z_2 = a_2 + b_2 i$, then

C. Selected Answers

$z_1 z_2 = (a_1 a_2 - b_1 b_2) + (a_1 b_2 + a_2 b_1)i$ so

$$\begin{aligned}|z_1|^2|z_2|^2 &= (a_1^2+b_1^2)(a_2^2+b_2^2)\\ &= a_1^2 a_2^2 + a_1^2 b_2^2 + b_1^2 a_2^2 + b_1^2 b_2^2\\ &= (a_1^2 a_2^2 - 2a_1 a_2 b_1 b_2 + b_1^2 b_2^2) +\\ &\quad (a_1^2 b_2^2 + 2a_1 b_2 b_1 a_2 + b_1^2 a_2^2)\\ &= (a_1 a_2 - b_1 b_2)^2 + (a_1 b_2 + a_2 b_1)^2\\ |z_1|^2|z_2|^2 &= |z_1 z_2|^2\end{aligned}$$

where we have both added and subtracted $2a_1 a_2 b_1 b_2$ in the third line. This shows $|z_1||z_2| = \pm|z_1 z_2|$, and since both sides are positive, they are equal.

4. (b) $(1,0)$ (f) $(3\sqrt{2}, \frac{\pi}{4})$
   (k) $r = \sqrt{17}, \theta = \arctan(-\frac{1}{4})$
5. (a) 3 (e) $1+\sqrt{3}i$ (n) $-3\sqrt{3}-3i$
10. (a) For $Arg(zw) = Arg(z) + Arg(w)$, let $z = -i$ and $w = i$.
    For $Arg(zw) \neq Arg(z) + Arg(w)$, consider $z = -\frac{\sqrt{2}}{2} + i\frac{\sqrt{2}}{2}$ and $w = -\frac{\sqrt{2}}{2} + i\frac{\sqrt{2}}{2}$.

Exercises 2.4 ▲ Page 23

1. $e^{iy} = \sum_{n=0}^{\infty} \frac{(iy)^n}{n!}$

    $= 1 + iy + \frac{(iy)^2}{2!} + \frac{(iy)^3}{3!} + \frac{(iy)^4}{4!} + \ldots$

    $= \left(1 - \frac{y^2}{2!} + \frac{y^4}{4!} + \ldots\right) + i\left(y - \frac{y^3}{3!} + \ldots\right)$

    $= \cos y + i \sin y$

2. (a) 1 (e) $\frac{1}{2} + \frac{\sqrt{3}}{2}i$ (k) $-\frac{\sqrt{3}}{2} + \frac{1}{2}i$
   (p) $-\frac{\sqrt{2}}{2} - \frac{\sqrt{2}}{2}i$
4. (a) $re^{-i\alpha}$ (b) $re^{i\alpha} = r$
   (c) $re^{-i\alpha}$
8. (b) Since $e^{iz} = \cos z + i \sin z$ and $e^{-iz} = \cos z - i \sin z$, we have $e^{iz} + e^{-iz} = \cos z + i \sin z + \cos z - i \sin z = 2\cos z$, thus $\cos z = \frac{1}{2}(e^{iz} + e^{-iz})$.

Exercises 2.5 ▲ Page 26

1. (b) $z = e^{\frac{\pi}{2}i}$ (d) $z = \sqrt{2} e^{\frac{\pi}{4}i}$
2. (a) $\cos^2(\beta) - \sin^2(\beta)$ (b) $2\cos(\beta)\sin(\beta)$
   (c) $\cos^4(\beta) - 6\cos^2(\beta)\sin^2(\beta) + \sin^4(\beta)$
   (d) $4\cos^3(\beta)\sin(\beta) - 4\cos(\beta)\sin^3(\beta)$
3. (a) $\cos\theta \sin\beta + \sin\theta \cos\beta$
   (b) $\cos\theta \cos\beta - \sin\theta \sin\beta$
   (c) $\dfrac{\cos\theta \sin\beta + \sin\theta \cos\beta}{\cos\theta \cos\beta - \sin\theta \sin\beta}$

4. (a) $\sqrt{2}^{23}(\cos(\frac{23\pi}{4}) + i\sin(\frac{23\pi}{4}))$ (h) 1

Exercises 2.6 ▲ Page 29

1. (a) $1, -1$

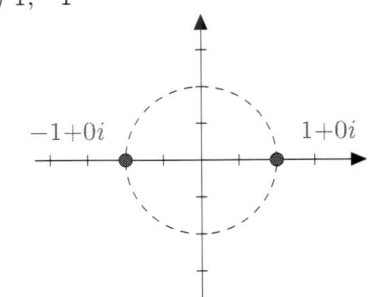

(c) $1, i, -1, -i$

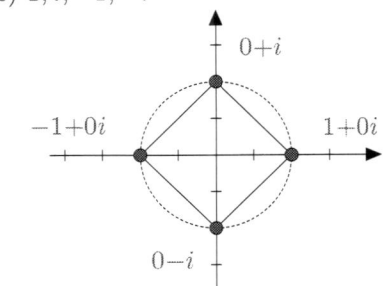

(e) $1, \frac{\sqrt{2}}{2} + \frac{\sqrt{2}}{2}i, i, -\frac{\sqrt{2}}{2} + \frac{\sqrt{2}}{2}i, -1, -\frac{\sqrt{2}}{2} - \frac{\sqrt{2}}{2}i, -i, \frac{\sqrt{2}}{2} - \frac{\sqrt{2}}{2}i$

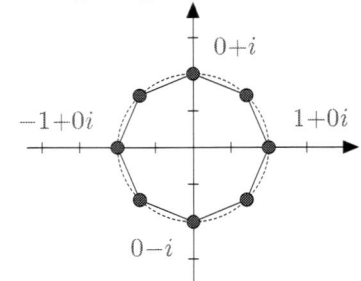

2. (a)

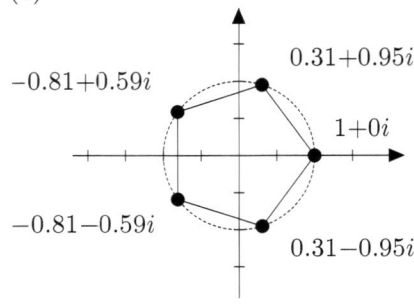

(c)

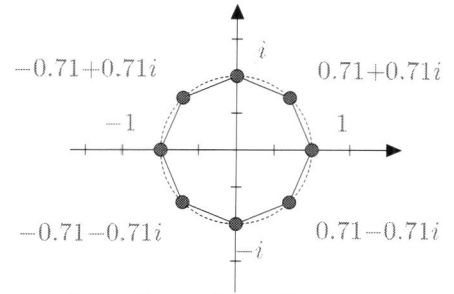

4. (b) $\frac{\sqrt{2}}{2}+\frac{\sqrt{2}}{2}i, -\frac{\sqrt{2}}{2}-\frac{\sqrt{2}}{2}i$
   (c) $2, 2i, -2, -2i$
   (f) $\sqrt{2}+\sqrt{2}i, -\sqrt{2}+\sqrt{2}i, -\sqrt{2}-\sqrt{2}i, \sqrt{2}-\sqrt{2}i$

Exercises 3.2 ▲ Page 36

1. We are given that $P(x) = Q(x)D(x) + R(x)$, $D(x) = x-c$, $°(R(x)) <° (D(x))$, and $°(x-c) = 1$, thus the degree of $R(x)$ must be zero, so $P(x) = Q(x)(x-c) + r$. Substituting c for x, $P(c) = Q(c)(c-c) + r = r$.

2. Given $x - c$ is a factor of the polynomial $P(x)$, we know that the remainder is 0. By the remainder theorem, $P(c) = 0$. Now changing directions, we are given that $P(c) = 0$, so $x - c$ is a factor of the polynomial $P(x)$.

7. (a) $(-2)^4 - 3(-2)^3 - 6(-2)^2 + 3(-2) - 10 = 0$
   (c) $(7)^3 - 6(7)^2 - 8(7) + 7 = 0$
   (f) $(2)^{103} - (2)^{102} - 4(2)^{100} = 0$

9. (a) -36 (d) $\frac{9}{2}$ (f) -4 (h) $i - 3$

10. (a) Quotient: x^2+6x+6; remainder: $-x-9$
    (c) Quotient: $2x^2 + 3x + 1$, remainder: 0

Exercises 3.3 ▲ Page 40

1. (a) Quotient: $3x^2 - 4x - 3$, remainder: 1
   (c) Quotient: $2x^5 + 2x^4 + 2x^3 + 2x^2 + 2x + 12$, remainder -25
   (f) Quotient: $x^3 - 3x^2 - 7x - 2$, remainder: 5

Exercises 3.4 ▲ Page 46

2. (a) $x^2 + 3$ (c) $-\frac{3}{4}x - \frac{3}{2}$
   (e) The polynomials are relatively prime.

Exercises 4.1 ▲ Page 53

3. (a) $P(x) = (x^3 - 3x^2 - 10x + 24)Q(x)$
   (c) $P(x) = (x^3 - 4x^2 + 14x - 20)Q(x)$
   (e) $P(x) = (x^3 + x^2 + x + 1)Q(x)$

4. (a) $-3, -2, 2$ (c) $-2, \frac{1}{2}, \frac{1}{2}, 1$

Exercises 4.2 ▲ Page 55

2. $P(x) = (x^8 - 4x^7 + x^6 + 12x^5 - 9x^4 - 12x^3 + 11x^2 + 4x - 4)Q(x)$

4. (a) $P(x) = (x-1)(x+\frac{1}{2}-\frac{\sqrt{3}}{2}i)(x+\frac{1}{2}+\frac{\sqrt{3}}{2}i)$
   (b) $P(x) = (x-1)(x+1)(x-i)(x+i)$
   (c) $P(x) = (x-i)(x+i)$
   (g) $P(x) = (x-\sqrt{-i})(x+\sqrt{-i})$

5. $(x^2 - 2x + 2)(x^2 + 2x + 2)$

Exercises 4.3 ▲ Page 58

2. (a) $P(x) = x^3 - 8x^2 + 29x - 52$
   (e) $P(x) = x^4 - \frac{2}{\sqrt{2}}x^3 + 2x^2 - \frac{2}{\sqrt{2}}x + 1$

3. (a) $2 + 3i, 2 - 3i, \sqrt{3}i, -\sqrt{3}i$
   (c) $1 - i, 1 + i, -1 - i, -1 + i$

Exercises 4.4 ▲ Page 61

1. (a) 2 (triple zero), -4 (double zero)
   (c) 2 (double zero), $1 - i, 1 + i$
   (e) 1 (double zero), $-1 - \sqrt{2}i, -1 + \sqrt{2}i$
   (h) $-1 - \sqrt{2}$ (double zero), $-1 + \sqrt{2}$ (double zero)

3. Consider $P(x) = x^3 + x^2 - x + 1$, then $P'(x) = 3x^2 + 2x - 1$. Notice that $P'(x)$ has a zero at -1 and $P(x)$ does not have a zero at -1.

Exercises 4.5 ▲ Page 63

2. (a) An example is exercise 4(c).

4. (a) $L(x) = \frac{1}{56}x^3 - \frac{7}{24}x^2 + \frac{11}{12}x - \frac{31}{21}$
   (c) $L(x) = 2x^2 - 1$

Exercises 5.1 ▲ Page 69

1. (a) Possible zeros: $\pm 12, \pm 6, \pm 4, \pm 3, \pm 2, \pm 1$; zeros: $-3, -2, 2$
   (c) Possible zeros: $\pm 6, \pm 3, \pm 2, \pm 1, \pm \frac{1}{2}, \pm \frac{1}{3}, \pm \frac{1}{6}, \pm \frac{2}{3}, 4 \pm \frac{3}{2}$; zeros: $\frac{2}{3}, 1, \frac{3}{2}$
   (e) Possible zeros: $\pm 18, \pm 9, \pm 6, \pm 3, \pm 2, \pm 1, \pm \frac{1}{3}, \pm \frac{2}{3}$; zeros: $-3, -2, \frac{1}{3}, 3$
   (g) Possible zeros: $\pm 28, \pm 14, \pm 7, \pm 4, \pm 2, \pm 1, \pm \frac{28}{3}, \pm \frac{28}{5}, \pm \frac{28}{15}, \pm \frac{14}{3}, \pm \frac{14}{5}, \pm \frac{14}{15}, \pm \frac{7}{3}, \pm \frac{7}{5}, \pm \frac{7}{15}, \pm \frac{4}{3}, \pm \frac{4}{5}, \pm \frac{4}{15}, \pm \frac{2}{3}, \pm \frac{2}{5}, \pm \frac{2}{15}, \pm \frac{1}{3}, \pm \frac{1}{5}, \pm \frac{1}{15}$; zeros: $\frac{-4}{3}, \frac{7}{5}$ are the only rational zeros.
   (i) Possible zeros: $\pm 2, \pm 1, \pm \frac{2}{9}, \pm \frac{1}{9}, \pm \frac{1}{3}, \pm \frac{1}{6}, \pm \frac{2}{3}, \pm \frac{1}{2}$; zeros: 1 is the only rational zero.

2. By the Rational Zero Theorem, we know that the only possible rational zeros of $x^2 - 6$ are $\pm 6, \pm 3, \pm 2, \pm 1$. Since none of these are zeros of the polynomial, we know that $\sqrt{6}$ is irrational.

3. We know that the golden ratio is the largest zero of $x^2 + x + 1$. By the Rational Zero Theorem, we have that the only possible rational zeros of the polynomial are ± 1. Since

C. Selected Answers

neither of these are zeros of $x^2 + x + 1$, the golden ratio must be irrational.

Exercises 5.2 ▲ Page 74

1. (a) Possible zeros: $\pm 12, \pm 6, \pm 4, \pm 3, \pm 2, \pm 1$; zeros: $-3, -2, 2$
   (c) Possible zeros: $\pm 48, \pm 24, \pm 16, \pm 12, \pm 8, \pm 6, \pm 4, \pm 3, \pm 2, \pm 1$; zeros: $-2, 1, 2, 12$
2. (a) There are no rational zeros.
   (c) The only rational zero is 13.

Exercises 5.3 ▲ Page 78

2. (a) We apply Eisenstein's Criterion with $q = 5$. Notice that 5 divides every coefficient except the first one and 5^2 does not divide the constant term. By Eisenstein's Criterion, $x^4 + 15x^3 + 35x^2 - 60x + 205$ is irreducible.
   (c) We may apply Eisenstein's Criterion with $q = 4$. Notice that 4 divides every coefficient except the first one and 4^2 does not divide the constant term. By Eisenstein's Criterion, $6x^8 + 256x^6 + 356x^3 - 60x + 52$ is irreducible.
   (f) First apply Eisenstein's Criterion with $q = 3$. Notice that 3 divides every coefficient except the first one and 3^2 does not divide the constant term. By Eisenstein's Criterion, $2x^7 + 3x^5 - 18x^2 + 6x + 30$ is irreducible.
   (i) We may apply Eisenstein's Criterion with $q = 7$. Notice that 7 divides every coefficient except the first one and 7^2 does not divide the constant term. By Eisenstein's Criterion, $4x^5 - 14x^4 - 91x^3 + 98x - 21$ is irreducible.
5. Given $P(x) = x^3 + 3x + (9k + 2)$ let $Q(x) = P(x+1)$, then $Q(x) = x^3 + 3x^2 + 6x + 3(3k+2)$. Now by applying Eisenstein's Criterion with $q = 3$, we get that $Q(x)$ is irreducible. Thus $P(x)$ is irreducible.
6. Given $P(x) = x^4 + 2x^2 + (4k - 1)$ let $Q(x) = P(x + 1)$, then $Q(x) = x^4 + 4x^3 + 8x^2 + 8x + 2(2k + 1)$. Now by applying Eisenstein's Criterion with $q = 2$, we get that $Q(x)$ is irreducible. Therefore $P(x)$ is irreducible.

Exercises 5.4 ▲ Page 83

4. (a) $-1, i, -i$

(l) $-1, -1, \frac{-\sqrt{15}}{4}i - \frac{1}{4}, \frac{\sqrt{15}}{4}i - \frac{1}{4}$
(o) $-1, -i, i, \frac{-1}{2} + \frac{\sqrt{3}}{2}i, \frac{-1}{2} - \frac{\sqrt{3}}{2}i$

Exercises 6.1 ▲ Page 98

1. (a) 4; 3; -12; -12 (c) -1; 0; -3; -3
   (f) 0; 0; 0; $-1,000,000$
   (h) 0; -1; There are only 2 zeros; -1
2. (a) $N = 20$; $r_1 = -2$; $r_2 = 2$; $r_3 = 5$
   (c) $N = 1$; $r_1 = 2 - 3i$; $r_2 = 2 + 3i$; $r_3 = -3$
   (g) $N = -4$; $M = 6$; $r_1 = 1 + i$; $r_2 = 1 - i$; $r_3 = -3$
   (j) $N = 9$; $r_1 = 3i$; $r_2 = -3i$; $r_3 = 7$
12. $x^3 - 2x^2 - 5x + 26$ has zeros $-2, 2 + 3i$, and $2 - 3i$.

Exercises 6.2 ▲ Page 104

5. (a) $Q(x) = x^3 + x^2 - 3x + 2$
   (c) $Q(x) = x^3 + 10x^2 + 30x + 29$
   (f) $Q(x) = x^3 + 20x^2 + 50x + 125$
   (h) $Q(x) = x^3 + 2x^2 + 4x + 1$
   (l) $Q(x) = x^3 + 5x^2 - 5x + 7$
   (n) $Q(x) = x^3 + \frac{7}{2}x^2 + \frac{13}{4}x + 1$
   (r) $Q(x) = x^3 + x^2 + 3x - 3$
6. (b) $Q(x) = 2x^3 - 30x^2 + 154x - 263$
   (e) $Q(x) = 2x^3 + 16x + 56$
   (i) $Q(x) = 7x^3 + 12x^2 + 54$
   (m) $Q(x) = 2x^3 + 12x^2 + 60x + 277$
   (q) $Q(x) = -31x^3 + 28x^2 - 12x + 2$
   (t) $Q(x) = 7x^3 - 30x^2 + 36x + 46$
8. (a) $c = -2$; $Q(x) = x^3 - 14x + 21$
   (c) $c = -\frac{1}{2}$; $Q(x) = 2x^3 - \frac{21}{2}x + 5$

Exercises 6.3 ▲ Page 107

1. (a) $Q(x) = x^2 + 2x + 1$ (c) $Q(x) = x^2 + x + 1$
   (e) $Q(x) = x^4 + x^3 + x^2 + x + 1$
2. $Q(x) = a^2x^2 + (2a - b^2)x + c^2$
7. Refer to Section 6.2

Exercises 6.4 ▲ Page 111

2. (e) This is symmetric since it is unchanged by interchanging any pair of variables.
   (b) Symmetric   (c) Symmetric
   (g) Not symmetric because when it is multiplied out it has the term $2x_5^2$, which differs from $3x_1^2$.
   (h) Not symmetric since it has terms $x_1 x_2^3 x_3$ and $x_1 x_2 x_3^3$, yet lacks $x_1^3 x_2 x_3$.
3. (a) $f_1^2 - 4f_2$

Exercises 7.3 ▲ Page 128

3. (a) 1.305407289, 5.758770483, −1.064177772
   (c) .7983603243, 3.330058740, −1.128419064
   (e) 2.254101688, 3.860805853,
   −0.1149075415

Exercises 7.4 ▲ Page 133

1. (a) $2 + \sqrt{6}, 2 - \sqrt{6}, 1 - \sqrt{5}i, 1 + \sqrt{5}i$
   (c) $-1 - \sqrt{3}, -1 + \sqrt{3}, 1 - \sqrt{6}, 1 + \sqrt{6}$
   (e) $-3, -2, 1, 2$ (g) $-2, 2, -1 + \sqrt{7}, -1 - \sqrt{7}$
   (i) $-1 - \sqrt{2}, -1 + \sqrt{2}, 1 - \sqrt{2}i, 1 + \sqrt{2}i$

Exercises 8.1 ▲ Page 139

1. Postulates 1 and 2 use a straight edge. Postulates 3, 4, and 5 use a compass.
3. (1) Archimedes Trisection with a marked ruler, see Exercise 7 in Section 8.6.
   (2) Tomahawk Trisection, see Exercise 10 in Section 8.6.

Exercises 8.2 ▲ Page 143

1. Construct a circle with center P_1 and radius $\overline{P_1P_2}$. Now construct a circle with center P_2 and radius $\overline{P_1P_2}$. Label the intersection points of the circles A and B. Draw line segments $\overline{P_1A}$ and $\overline{P_2A}$. Now draw the line segment $\overline{AB}$. Label the intersection of $\overline{AB}$ $\overline{P_1P_2}$, P_3. P_3 is the midpoint of $\overline{P_1P_2}$ and thus $\overline{AB}$ bisects $\overline{P_1P_2}$.
2. Given an angle ∠BAC, let D be an any point on $\overline{AB}$. Construct E on $\overline{AC}$ with $AD = AE$. Now Construct a circle with center D and radius $\overline{AD}$. Construct another circle with center E and radius $\overline{AE}$. The two circles intersect at point A and another point, call it F. Construct the ray $\overrightarrow{AF}$, which bisects ∠BAC.
3. (a) Let the endpoints of $\overline{x}$ be A and B. Since the lengths x and y are constructible, copy the length of y and add that to x at point B. Then $x + y$ is constructible.
   (b) We know x and y are constructible. Let $x > y$. Let the endpoints of $\overline{x}$ be A and B. Now copy the length y and measure this length from point B toward point A. Label this point C, on $\overline{x}$. Then $\overline{AC}$ has the length $x - y$.
4. (a) Given two points A and B, construct a circle with center A and radius $\overline{AB}$. Now construct a circle with center B and radius $\overline{AB}$. Label the intersection points of the circles C and D. Draw line segments $\overline{AC}$ and $\overline{BC}$. △ABC is an equilateral triangle.

Exercises 8.6 ▲ Page 165

1. (a) To construct an equilateral triangle, given two points A and B, begin by drawing two circles. One with center A and radius $\overline{AB}$ and the other with center B and radius $\overline{AB}$. Let the intersection points of the circles be C and D. Draw lines $\overline{AC}$ and $\overline{BC}$. Now we have △ABC. Notice that we drew 2 circles and 2 lines. Then the total cost is 30 cents, which is less than 50 cents.

Exercises 9.1 ▲ Page 178

1. (a) $(-\infty, -\sqrt{31/22}), (-\sqrt{31/22}, 0),$ $(0, \sqrt{31/22}), (\sqrt{31/22}, \infty)$
   (b) $(-\infty, -\sqrt{2}), (-\sqrt{2}, \sqrt{2}), (\sqrt{2}, \infty)$
   (c) $(-\infty, -\sqrt{5}), (-\sqrt{5}, 0), (0, \sqrt{5}), (\sqrt{5}, \infty)$
3. Consider the polynomial $P(x)$ with consecutive real zeros a and b, meaning $P(a) = 0$ and $P(b) = 0$. Let $b > a$. Then by the Mean Value Theorem, $P'(c) = \frac{P(b)-P(a)}{b-a} = \frac{0-0}{b-a} = 0$, for $b > c > a$. Thus there is at least one zero of $P'(x)$ between every two consecutive real zeros of $P(x)$.

Bibliography

[1] M. Abramowitz and I. Stegun. *Handbook of Mathematical Functions,* New York: Dover Publications, 1972. ISBN: 978-0486612720. This once was the classic reference with descriptions and tables of values for mathematical functions. The internet and computer algebra systems have removed most of the need for this text, but it is still a fine reference.

[2] D. J. Acosta, *Newton's Rule of Signs for Imaginary Roots,* The American Mathematical Monthly, Vol. 110, **8**, Oct. 2003, pp. 694–706.

[3] A. A. Albert. *An Inductive Proof of Descartes' Rule of Signs,* The American Mathematical Monthly, Vol. 50, **3**, Mar. 1943, pp. 178–180.

[4] D. Kalman, *Uncommon Mathematical Excursions: Polynomia and Related Realms*, American Mathematical Society, 2009. ISBN 978-0-88385-341-2. A wonderful text describing polynomials and related material.

[5] M. E. Lill, "Résolution graphique des équations numériques de tous degrés à une seule inconnue, et description d'un instrument inventé dans ce but," Nouvelles Annales de Mathmatiques, 2:6, 1867, pp 359–362. The original article describing Lill's method.

[6] M. E. Lill, "Résolution graphique des équations algébriques qui ont des racines imaginaires," Nouvelles Annales de Mathématiques, 2:7, 1868, pp 363–367.

[7] H. Abe, Possibility of trisection of arbitrary angle by paper olding in SUGAKU Seminar (in Japanese), Suken ublishing, Kyoto, 1980. The first article describing how to trisect angles with origami.

[8] E. T. Bell. *Men of Mathematics,* New York: Simon and Schuster, 1965. Reissued by Touchstone Book in 1986, ISBN: 978-0671628185. A well-known and often quoted history of mathematics. Very witty and engaging.

BIBLIOGRAPHY

[9] W. Bishop. *How to Construct a Regular Polygon,* The American Mathematical Monthly, Vol 85, **3**, pp 186-88. This article gives a tower of quadratic polynomials to be solved to construct the 257-gon, and a general procedure for finding such polynomials. Requires a knowledge of Galois theory.

[10] J. M. Borwein & P. B. Borwein, *Irrationality Measures,* section 11.3 of "Pi & the AGM: A Study in Analytic Number Theory and Computational Complexity." New York: Wiley, pp. 362-386, 1987. ISBN: 978-0471831389.

[11] N. B. Conkwright. 1957. *Introduction to the Theory of Equations,* Ginn and Company, 1972. No longer in print. Concise.

[12] J. H. Conway. *On Numbers and Games,* 2nd edition, A. K. Peters/CRC Press; 2nd edition, 2000. ISBN: 978-1568811277. This book starts with a beautiful definition of (finite and transfinite) numbers which is generalized to include a large class of games.

[13] D. E. Dobbs and R. Hanks. *A Modern Course on the Theory of Equations,* 2nd edition, Polygonal Publishing House, 1992. ISBN: 978-0936428147. This book attempts to develop the theory of equations along with field theory, in an informal conversational historical style. It fails: the conversational style makes it difficult to follow and hard to find the theorems.

[14] U. Dudley, *The Trisectors,* Spectrum, Mathematical Association of America, 1996. ISBN: 978-0883855140.

[15] Euclid (translated by T. L. Heath). *The Thirteen Books of Euclid's Elements* 3 volumes. Dover Publications, 1956. One of the greatest classics of ancient mathematics.

[16] A. O. Gelfond (translated by L. Boron). *Transcendental and Algebraic Numbers.* New York: Dover Publications, 1960. This is an excellent book on the topic of the title—too difficult for the average undergraduate.

[17] M. J. Greenberg. *Euclidean and Non-Euclidean Geometries,* San Francisco: Freeman and Company, 1974. Many interesting historical notes.

[18] E. Grosswald. *Recent Applications of Some Old Work of Laguerre.* American Mathematical Monthly, Vol 86, **8**, Oct. 1979, pp 648–658. Discusses the center of mass of the zeros of a polynomial; a condition for the roots of a polynomial to all be real; and bounds for roots of certain classical polynomials. Advanced undergraduate level.

[19] G. H. Hardy and E. M. Wright. *An Introduction to the Theory of Numbers,* 6th edition, Oxford University Press, 2008. ISBN: 978-0199219865. One of the

true classics in number theory. Everyone interested in this subject should own a copy.

[20] T. Heath. *A History of Greek Mathematics.*

[21] E. W. Hobson, et al. *Squaring the Circle,* New York: Chelsea Publishing, 1969. This nice little book is actually four small books found and republished together by Chelsea Publishing. The first is "Squaring the Circle, a History of the Problem," a well detailed 57 page history, including related subjects such as approximating π with ruler and compass, and giving a proof of the transcendence of π. The second is "Ruler and Compasses", a 143-page work discussing which constructions are possible with ruler and compass; compass alone; and ruler alone.

[22] T. Hull. *Project Origami: Activities for Exploring Matrhematics,* A. K. Peters, MA, 2006. ISBN 1-56881-258-2. A wonderful book of discovery-based classroom learning activities which show how origami can be used to teach many areas of mathematics.

[23] T. Hull. "Solving cubics with creases: the work of Beloch and Lill," American Mathematical Monthly **118**, April 2011, 307–315. Shows how to solve the general cubic using an origami version of Lill's method.

[24] F. Klein *et al. Famous Problems and other Monographs.* New York: Chelsea, 1980. The first part of this collection deals with geometric constructions and contains proofs that e and π are transcendental.

[25] W. V. Lovitt. *Elementary Theory of Equations,* Prentice-Hall, 1939. No longer in print. Of interest because of a reasonable and short treatment of ruler and compass constructions.

[26] J. W. Pratt. *Finding How Many Roots a Polynomial has in $(0,1)$ or $(0,\infty)$.* The American Mathematical Monthly, Vol 86, **8**, Oct 1979, pp 630–637. Provides an easy alternative to Sturm's theorem for finding the number of roots in a given interval.

[27] D. Shanks. *Solved and Unsolved Problems in Number Theory.* New York: Chelsea, 1978. Reprinted by AMS Chelsea Publishing in 2002. ISBN: 978-0821828243. A very good reference for the more advanced reader.

[28] M. R. Spiegel. *Schaum's Outline of Theory and Problems of College Algebra,* 4th edition, Schaum's Outline Series. McGraw-Hill, 2010. ISBN: 978-0071626477. A concise summary of the elementary results of chapters 1, 2 and 3 of these notes. Of interest because of its many worked examples. Available in most college book stores.

Bibliography

[29] P. A. Stark. *Introduction to Numerical Methods,* 2nd edition, Macmillan Publishing, New York, 1970. ISBN: 978-0024160706. Most any other book of the same title is as good.

[30] J. V. Uspensky. *Theory of Equations,* McGraw-Hill, 1963. ISBN-13: 978-0070667365. No longer in print. This is one of the best references on the theory of equations, definitely worth owning.

[31] B. L. Van Der Waerden (translated by F. Blum). *Modern Algebra,* New York: Frederick Ungar Publishing, 1953. No longer in print.

[32] R. C. Yates. *The Trisection Problem,* Classics in Mathematics Education, Vol 3. The National Council of Teachers of Mathematics, 1971. ISBN-13: 978-0873530385. This is a short (68 pages) work concerned solely with trisection including: the problems history; which angles are trisectable; solutions using curves and mechanical devices; approximations; and a note on those giving false solutions. A very good reference on this problem.

Index

$\mathbb{C}$, 8, 12
$\Im(z)$, 14
$\Re(z)$, 14
Var(P, a), 193
Var$(\delta P, x)$, 248
arg(z), 18
R, 7
🖥, ix
📁, ix
(N), ix
(A), ix
(*), ix
(**), ix
(***), ix

absolute value, 12
algebraic numbers, 231, 255
algebraic solution, 119
algebraically closed field, 56
amplitude, 16
analytic function, 56, 246
Arg(z), 20
argument, 16, 18
argument, principle, 18, 20

biquadratic, 34

Cardan, 10
Cardan's solution, 122, 124
carpenter's square, 167
Cauchy bound, 183, 184

cis α, 23
coefficient, 33
coefficients, 95
common divisor, 42
common factor, 42
complex conjugate, 8
complex number, 8
composite number, 75
conjugate zero, 57
constant term, 33
constructible by ruler & compass, 141
construction, geometric, 141
cubic, 34
cubic resolvent, 130
cubic, reduced (form), 122
cubic, solution, 122
cubic, trig. solution, 127

De Moivre's Theorem, 24, 243
degree, 33
degree of a polynomial, 33
degree, alg. number, 231, 232, 234
degree, polynomial, 120
Descartes' Rule of Signs, 179
discriminant, 255
discriminant, cubic, 122, 126–128, 178, 254
discriminant, quadratic, 95, 121, 254
divides, 36
divides, polynomial, 36

Index

Division Algorithm, 35

Eisenstein's Criterion, 76, 78, 79
elementary sym. functions, 109
equivalent conditions, 80
equivalent definitions, 111
Euclidean Algorithm, 43, 193, 247
Euler ϕ-function, 30
Euler's Formula, 21
Euler's Theorem, 229, 230
Extended Factor Theorem, 51

Factor Theorem, 36
factor, polynomial, 36
factorial, 75
False Position, 206
Fermat, 154
Fermat number, 155
Fermat prime, 138, 155
Ferrari, 125
Fundamental Theorem of Algebra, 54, 243
Fundamental Theorem on Symmetric Functions, 109

Gauss, 154
gcd, polynomial, 42, 46, 62, 193, 253, 255
general equation, 120
golden ratio, 70, 234
greatest common divisor, 42, 46, 60, 62, 193, 253, 255
greatest common measure, 46

highest common factor, 46
Horner's Process, 101
Horner's Rule, 41

identical, 33
identically vanishing, 34
Identity Theorem, 53
imaginary axis, 12
imaginary number, 10

imaginary part, 8
integral, 71
integral polynomial, 34
Integral Zero Theorem, 71
Intermediate Value Theorem, 205
irrational, 67
irrational number, 221
irrationality measure, 235
irreducible case, 122
irreducible modulo m, 79
irreducible polynomial, 75, 76, 78, 152, 228
iteration, simple, 214

Lagrange bound, 186
Lagrangian Interpolation, 62
leading coefficient, 33
leading term, 33
Lill's method, 85
Lindemann-Weierstrass Theorem, 237
linear, 34
Liouville number, 235, 237
Liouville's constant, 234
Liouville's Theorem, 56, 232, 246

mean value theorem, 178
modulus, 12, 16
monic polynomial, 34
monomial, 33
multiplicity, 55, 60

Newton's Identities, 113
Newton's Incomplete Rule, 180
Newton's Method, 207
Newton-Raphson, 207
non-zero polynomial, 34
normal form, 8
number, composite, 75
number, decimal expansion, 222
number, irrational, 67, 221
number, number, vi, 152, 231, 232
number, prime, 75
number, rational, 67, 221, 231

270

number, terminating expansion, 222

origami, 87
origami geometry, 164

partial fraction decomposition, 64
Pepin's Test, 155
period, 222
polar form, 16
polynomial, 33
polynomial equation, 51
polynomial, divisor, 36
polynomial, factor, 36
polynomial, integral, 34
polynomial, irreducible, 75, 76, 78, 152, 228
polynomial, monic, 34
polynomial, rational, 34
polynomial, reciprocal, 80
polynomial, reducible, 75
prime number, 75
primitive root, 29
principle argument, 18, 20
pure imaginary, 8

quadrant, 18
quadratic, 34
quadratic formula, 120
quartic, 34
quintic, 34
quotient, 35

rational, 67
rational number, 221, 231
rational polynomial, 34
Rational Zero Theorem, 67, 222
real axis, 12
real number, 7
real number line, 12
real part, 8
reciprocal polynomial, 80, 84
rectangular form, 12
rectify a circle, 153

reducible modulo m, 79
reducible polynomial, 75
Regula Falsi, 206
regular polygon, 29
remainder, 35
Remainder Theorem, 36
resolvent, 130
resultant, 253
ring, polynomial, 35
root, 51
roots of unity, 28
Roth's Theorem, 236

Scipione, 124
secant method, 214
simple iteration, 214
solution by radicals, 119
solvable, 119
Sturm sequence, 193
Sylvester matrix, 254
symmetric function, 108, 253
symmetric polynomial, 108, 111
synthetic division algorithm, 38

Tartaglia, 125
terms, 33
tomahawk trisector, 168
transcendental number, vi, 152, 231, 232
transforming, 100
translation, 100
triangle inequality, 14
trisection equation, 129

variable, 33

Weierstrass' Nullstellensatz, 201, 206

zero polynomial, 34
zero, conjugate, 57
zero, double, 55, 60
zero, integral, 71
zero, multiple, 60, 71, 112, 126, 255

Index

zero, of a polynomial, 51
zero, rational, 67, 70, 124, 186, 222
zero, simple, 55, 60
zero, triple, 55, 60
zeros, transforming, 100
zeros, translation, 100

Printed in Great Britain
by Amazon